DURABILITY OF STRUCTURAL ADHESIVES

SEVEN DAY LOAN

This book is to be returned on
or before the date stamped below

1 8 FEB 2002	
2 5 FEB 2002	
3 0 MAY 2002	
2 8 JAN 2004	

DURABILITY OF STRUCTURAL ADHESIVES

Edited by

A. J. KINLOCH

BSc, PhD, C Chem, FRSC, FPRI

Ministry of Defence, Procurement Executive, Propellants, Explosives and Rocket Motor Establishment, Waltham Abbey, Essex, UK

APPLIED SCIENCE PUBLISHERS
LONDON and NEW YORK

APPLIED SCIENCE PUBLISHERS LTD
Ripple Road, Barking, Essex, England

Sole Distributor in the USA and Canada
ELSEVIER SCIENCE PUBLISHING CO., INC.,
52 Vanderbilt Avenue, New York, NY 10017, USA

British Library Cataloguing in Publication Data

Durability of structural adhesives.
1. Adhesives
I. Kinloch, A. J.
668'.3   TP968

ISBN 0-85334-214-8 ✓

WITH 31 TABLES AND 142 ILLUSTRATIONS

© APPLIED SCIENCE PUBLISHERS LTD 1983

The selection and presentation of material and the opinions expressed in this publication
are the sole responsibility of the authors concerned.

Printed in Northern Ireland at The Universities Press (Belfast) Ltd.

Preface

As a means of joining materials the use of adhesives offers considerable advantages when compared to other, more conventional methods, such as brazing, welding, riveting, bolting, etc., and the many advantages have led to a continuing increase in the use of adhesives and a widespread appreciation of the benefits of adhesives technology. These comments are particularly apposite in the case of structural adhesive bonding.

There is no universally accepted definition of a structural adhesive but the present text considers such adhesives to be based upon monomer compositions which polymerise to give high-modulus, high-strength adhesives between rigid adherends, so that a load-bearing joint is constructed. Structural adhesives are used extensively in the aerospace industry to join metals, such as aluminium, steel, titanium and their respective alloys, and increasingly fibre-laminate adherends, such as carbon-fibre and glass-fibre reinforced plastics. However, they are also widely used in the general engineering and construction industries.

One of the most important requirements of a structural adhesive joint is the ability to retain a significant proportion of its load-bearing capability for long periods under the wide variety of environmental conditions which are likely to be encountered during its service life. Unfortunately, one of the most hostile environments for structural adhesive joints is water and this is, of course, one of the most commonly encountered. Indeed, by far the most important problem currently facing adhesive scientists and technologists is that of ensuring the long-term durability of structural adhesive joints exposed to environments where the concentration of water is relatively high. Thus, the present book is devoted to a consideration of the durability of structural adhesives and largely focuses upon environmental attack by moisture.

A. J. KINLOCH

v

Contents

List of Contributors

P. ALBERICCI

British Aerospace, Manchester Division, Chester Road, Woodford, Stockport, Cheshire, UK.

D. M. BREWIS

School of Chemistry, Leicester Polytechnic, PO Box 143, Leicester LE1 9BH, UK.

W. BROCKMANN

Fraunhofer-Institut für Angewandte Materialforschung, Lesumer Heerstrasse 36, D-2820 Bremen 77, Federal Republic of Germany.

J. COMYN

School of Chemistry, Leicester Polytechnic, PO Box 143, Leicester LE1 9BH, UK.

G. D. DAVIS

Materials Department, Martin Marietta Laboratories, 1450 South Rolling Road, Baltimore, Maryland 21227, USA.

A. J. KINLOCH

Ministry of Defence, Procurement Executive, Propellants, Explosives and Rocket Motor Establishment, Waltham Abbey, Essex EN9 1BP, UK.

A. MAHOON

British Aerospace, Brooklands Road, Weybridge, Surrey, UK.

J. D. MINFORD

Aluminum Company of America, Product and Process Engineering Division, Alcoa Technical Center, Alcoa Center, Pennsylvania 15069, USA.

J. D. VENABLES

Materials Department, Martin Marietta Laboratories, 1450 South Rolling Road, Baltimore, Maryland 21227, USA.

1

Introduction

A. J. KINLOCH

Propellants, Explosives and Rocket Motor Establishment, Waltham Abbey, UK

NOTATION

a crack length
p vapour pressure
p_0 equilibrium vapour pressure
ΔH_v molar heat of vaporisation
R   gas constant
T temperature
V molar volume
W_A thermodynamic work of adhesion in dry air
W_{AL} thermodynamic work of adhesion in a liquid environment
γ surface free energy
γ^D dispersion force component to the surface free energy
γ^P polar force component to the surface free energy
γ_{Ad} γ for cured adhesive
γ_{AdL} γ for cured adhesive/liquid interface
γ_c critical surface tension of wetting
γ_{LV} γ for test liquid or liquid adhesive
γ_S γ for substrate or adherend
γ_{SAd} γ for substrate (or adherend)/cured adhesive interface
γ_{SL} γ for substrate (or adherend)/liquid interface
γ_{SV} γ for substrate (or adherend)/vapour interface
δ solubility parameter
θ contact angle
π_S spreading pressure
Γ surface concentration of adsorbed vapour

1. DEFINITIONS, BACKGROUND AND SCOPE

An *adhesive* may be defined as a material which when applied to surfaces of materials can join them together and resist separation. Adhesive is the general term and includes cement, glue, paste, etc. The term *adhesion* is used when referring to the attraction between the substances. The materials being joined are commonly referred to as the *substrates* or *adherends*, and the latter term is particularly convenient when the materials are part of a joint.

As a means of joining materials the use of adhesive offers many advantages when compared to other, more conventional, methods such as brazing, welding, riveting, bolting, etc. The advantages include:

(a) the ability to join dissimilar materials, for example, metals to plastics and rubbers,

(b) the ability to join thin sheet material efficiently,

(c) an improved stress distribution in the joint which imparts, for example, an increase in fatigue resistance to the bonded component,

(d) an increase in the design flexibility,

(e) the fact that, frequently, it is the most convenient and cost-effective technique.

These advantages have led to a continuing increase in the use of adhesives and a widespread appreciation of the benefits of adhesives technology. These comments are particularly apposite in the case of structural adhesive bonding. There is no universally accepted definition of a *structural adhesive*, but in the present text we will consider such adhesives to be based upon monomer compositions which polymerise to give high-modulus, high-strength adhesives between relatively rigid adherends, so that a load-bearing joint is constructed. Structural adhesives are extensively used in the aerospace industry to join metals such as aluminium, titanium and their respective alloys and, increasingly, fibre-laminate adherends, such as carbon-fibre (cfrp) and glass-fibre (grp) reinforced plastics. However, they are also widely used in the general engineering and construction industries where, apart from the adherends listed above, wood and concrete are also frequently bonded using such adhesives.

Structural adhesives are often based upon low molar-mass phenolic or epoxy resins, and more recently acrylic resins, which polymerise, or cure (this process is also sometimes known as hardening), to give highly crosslinked adhesives, i.e. thermosetting polymers, which have chemical bonds connecting the polymer chains. These resins have gained wide acceptance for several reasons. For example, the resins initially possess a sufficiently low viscosity (although in some instances heat may be required) to flow over a substrate surface without the need to employ solvents. They are polar materials which (i) assist in removing atmospheric contamination which is invariably present on metallic substrates, and (ii) increase the degree of intrinsic adhesion (see ref. 1 and discussion below). Also, a variety of curing agents may be used, particularly in the case of epoxy resins, to give a wide range of possible times and temperatures for the curing reaction. Upon curing, these resins possess a high degree of crosslinking and a high modulus, high strength, low creep and good elevated-temperature properties. However, this chemical structure often produces a very brittle adhesive exhibiting a poor crack resistance. Therefore, it is usually necessary to increase the toughness of such adhesives if they are to be successfully used in structural applications and this is frequently achieved by the controlled inclusion of a rubber, which phase-separates when the resin is cured, to give a two-phase microstructure. This microstructure enables the good physical properties resulting from the high crosslink density in the cured resin phase to be maintained whilst the presence of the dispersed rubbery second phase greatly increases the toughness and peel strength of the adhesive. Since structural adhesives formulations also usually contain viscosity-control agents, corrosion-resistant fillers, fillers to improve heat resistance, etc., it may be appreciated that they are complex materials.

Some typical physical properties of a modern rubber-toughened structural adhesive, based upon an epoxy resin cured with dicyandiamide ('dicy'), are shown in Table 1. There are several noteworthy features. Firstly, all the joints failed by cohesive fracture through the adhesive, except for the cfrp–steel double-lap joints which failed by interlaminar fracture in the cfrp adherend. Thus, in these unaged joints, interfacial failure between the adherend and the adhesive was never recorded. This is a common observation and, indeed, an interfacial locus of joint failure is usually found only *after* environmental attack. In such instances a visual assessment of the fracture surfaces of the broken joint invariably suggests that the adherend has come clearly away from the adhesive, or from the primer if one has been employed. This interfacial failure mode resulting from environmental attack highlights the importance of the interfacial regions in any study of joint

TABLE 1

REPRESENTATIVE PHYSICAL PROPERTIES OF A RUBBER-TOUGHENED EPOXY-BASED STRUCTURAL ADHESIVE MEASURED IN BULK AND IN VARIOUS ADHESIVE JOINTS[2]

Test[a]	Adherends	Property	
Bulk adhesive			
Flexure	—	Modulus	2·8 GPa
	—	Failure stress	74·5 MPa
	—	Failure strain	2·7%
	—	Glass transition temperature	120°C
Adhesive joints[b,c]			
Torsional shear	Aluminium alloy	Failure stress	61 MPa
Axially-loaded butt joints	Steel	Failure stress	58 MPa
Single-lap joint, loaded in tension	Aluminium alloy	Failure load	9 kN
		Failure stress	28 MPa
	Steel	Failure load	12·3 kN
		Failure stress	38 MPa
Double-lap joint, loaded in tension	Cfrp–steel	Failure load	24 kN
		Failure stress	6·9 MPa
	Cfrp–steel; tapered adherends	Failure load	80 kN
		Failure stress	19·7 MPa
Peel tests			
90° peel	Aluminium alloy	Peel strength	5 kN/m
	Steel	Peel strength	0·6 kN/m
135° peel	Aluminium alloy	Peel strength	4 kN/m
Precracked, TDCB[d]	Steel	Fracture energy, G_{Ic}	0·9 kJ/m^2
Precracked, compact shear	Steel	Fracture energy, G_{IIc}	2·2 kJ/m^2

[a] Tests conducted at 23°C and a moderate rate.
[b] See Fig. 1.
[c] All joints failed by cohesive fracture through the adhesive, except for the cfrp–steel double-lap joints which failed in the cfrp adherend.
[d] TDCB: tapered-double-cantilever-beam specimen.

durability and this will be a recurring theme in the present and following Chapters.

Secondly, it may be seen from the results shown in Table 1 that very high joint strengths may be attained but the actual values are greatly dependent upon the detailed joint geometry and the adherend materials. This arises, of course, from the stress distributions in loaded adhesive joints generally being non-uniform. Thus, stress concentrations are generated and their magnitudes, and hence the measured

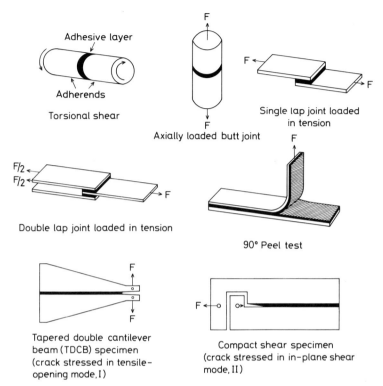

Fig. 1. Sketches of joint geometries employed to obtain the data shown in Table 1.

joint strength values, are highly dependent upon the detailed geometry and the mechanical properties of the adherends and the adhesive. The reader is referred elsewhere[2-5] for further detailed discussions on stress distributions in adhesive joints.

However, whilst the advantages of adhesives technology have been emphasised and the outstanding physical properties of modern structural adhesives have been indicated above, one of the most important requirements of a structural adhesive joint has yet to be discussed: namely, the ability to retain a significant proportion of its load-bearing capability for long periods under the wide variety of environmental conditions which are likely to be encountered during its service life. Unfortunately, one of the most hostile environments for structural adhesive joints is water and this, of course, is one of the most commonly encountered. Indeed, by far the most important problem currently facing adhesive scientists and technologists is that of the long-term durability of structural adhesive joints exposed to environments where the concentration of liquid water, or water vapour, is relatively high. Therefore, the present book is devoted to a consideration of the durability of structural adhesives and largely concentrates upon environmental attack by moisture.

This introductory chapter aims at providing a broad overview of the various aspects associated with the environmental failure of structural adhesive joints, thus to set the scene for the subsequent chapters in which the different features of the problem are discussed in considerable detail. These chapters are divided into two Parts. In Part I the fundamental aspects are reviewed. A recurring theme when discussing joint durability is the importance of the adhesive/adherend interface; the exciting recent developments in the analysis of surfaces and interfaces are reviewed in Chapter 2. This is followed by a discussion of the kinetics and mechanisms of environmental attack in Chapter 3. In Part II the materials which may form the joint are considered. In Chapter 4 the various types of structural adhesives which are generally commercially available are discussed whilst Chapters 5, 6 and 7 review the particular problems encountered when bonding aluminium, titanium and steel (and their different alloys) respectively. These particular adherends are considered in detail since, for the reasons discussed in this chapter and in Chapter 3, structural joints involving metallic adherends usually present the greatest problem to the adhesion scientist when long service life is required. Again, the importance of the adhesive/adherend interface is evident from the major influence the adherend surface pretreatment prior to bonding has upon the subsequent joint durability. Finally, Chapter 8 discusses the durability of structural adhesive joints which have been employed in the aerospace industries; the relatively long and successful history of bonded structures in aircraft leads to aerospace applications being of particular interest in a study of environmental aspects. It is hoped that the present book will enable the reader to gain a firm understanding of the fundamental science behind the environmental failure of structural adhesives and of the various methods, both theoretical and experimental, which have been employed to predict and prevent such failures.

The factors which affect the durability are outlined later in this chapter and the basic reasons why water is such an aggressive environment are considered. However, as considerable attention is focused in this book upon the adhesive/adherend interface, the reasons why substances do adhere and the nature of the intrinsic adhesion forces which act across the adhesive/adherend interface are first reviewed.

2. INTERFACIAL CONTACT AND INTRINSIC ADHESION

2.1. Interfacial Contact

It has been recognised for many years that the establishment of intimate molecular contact at the interface is a necessary, though sometimes insufficient, requirement for developing strong adhesive joints. This means that the adhesive needs to be able to be spread over the solid substrate, or adherend, surface and needs to displace air and any other contaminants that may be present on the surface. An adhesive which conforms ideally to these conditions must:

(a)   when liquid, exhibit a zero or near zero contact angle;
(b)   at some time during the bonding operation have a viscosity that should be relatively low, e.g. no more than a few centipoises;
(c)   be brought together with the adherends in a rate and manner that should assist the displacement of any trapped air.

In order to assess the ability of a given adhesive/adherend combination to meet these criteria it is necessary to consider wetting equilibria, to ascertain values of the surface free energies of the adhesive and substrate and the free energy of the adhesive/adherend interface and to examine the kinetics of the wetting process.

2.1.1. Wetting Equilibria

Surface tension is a direct measurement of intermolecular forces. The tension in surfaces layers is the result of the attraction of the bulk material for the surface layer and this attraction tends to reduce the number of molecules in the surface region resulting in an increase in intermolecular distance. This increase requires work to be done, and returns work to the system upon a return to a normal configuration. This explains why tension exists and why there is a surface free energy. The most common type of physical surface attractive forces are the van der Waals forces, which can be attributed to different effects: (a) dispersion (or London) forces arising from internal electron motions which are independent of dipole moments and (b) polar (or Keesom forces) arising from the orientation of permanent electric dipoles and the induction effect of permanent dipoles on polarisable molecules. The dispersion forces are usually weaker than the polar forces but they are universal and all materials exhibit them. Another type of force that may operate is the hydrogen bond, formed as a result of the attraction between a hydrogen atom and a second, small and strongly electronegative atom such as fluorine, oxygen or nitrogen.

Wetting may be quantitatively defined by reference to a liquid drop resting in equilibrium on a solid surface as shown in Fig. 2. The tensions at the three-phase contact point are indicated: subscript LV refers to the liquid/vapour point, SL to the solid/liquid point and SV to the solid/vapour point. The Young equation[6,7] relating these tensions to the equilibrium contact angle, θ, may be written as

$$\gamma_{SV} = \gamma_{SL} + \gamma_{LV} \cos \theta \tag{1}$$

The term γ_{SV} represents the surface free energy of the solid substrate resulting from adsorption of vapour from the liquid and may be considerably lower in value than the surface free energy of the solid in vacuo, γ_S. This reduction in the surface free energy of the solid when covered by a layer of vapour has been defined by the concept of equilibrium spreading pressure, π_S, such that when the vapour obeys the ideal gas law[8]

$$\pi_S = \gamma_S - \gamma_{SV} = \mathbf{R}T \int_0^{p_0} \Gamma \, d(\ln p) \tag{2}$$

where p is the vapour pressure, p_0 is the equilibrium vapour pressure, $\mathbf{R}$ is the gas constant, T is the absolute temperature and Γ is the surface concentration of the adsorbed vapour. Equation 1 may be

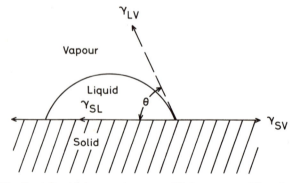

Fig. 2. *A liquid drop resting at equilibrium on a solid surface.*

rewritten as

$$\gamma_S = \gamma_{SL} + \gamma_{LV} \cos \theta + \pi_S \tag{3}$$

When $\theta > 0°$ the liquid adhesive is non-spreading but when $\theta = 0°$ the liquid wets the solid completely and spontaneously spreads freely over the surface at a rate depending upon the viscosity of the adhesive and the surface roughness of the solid substrate. Thus for spontaneous wetting to occur:

$$\gamma_{SV} \geq \gamma_{SL} + \gamma_{LV} \tag{4}$$

or

$$\gamma_S \geq \gamma_{SL} + \gamma_{LV} + \pi_S \tag{5}$$

It is also possible for a liquid to spread and wet a solid surface even when $\theta > 0°$ but this requires the application of a pressure or force to the liquid adhesive to spread it forcibly over the solid surface. However, before these concepts may be of use in adhesives technology the surfaces and the interfacial free energies need to be calculated and the kinetics considered.

2.1.2. Surface and interfacial free energies
A useful distinction may be made between low-energy and high-energy solid surfaces. Organic compounds, such as polymers, belong to the first group and their surface free energies are usually less than *ca* 100 mJ/m^2. Metals, metal oxides and ceramics belong to the second group since, when their surfaces are atomically clean, their surface free energies are typically greater than *ca* 500 mJ/m^2.

In the case of low-energy, polymeric solid surfaces the first approach adopted to their characterisation was in the 1950s when Zisman introduced the concept of 'critical surface tension'. This parameter was derived from the observation[9] that for low-energy solids and a series of homologous test liquids, e.g. hydrocarbons, a rectilinear relationship frequently existed between the cosine of the contact angle, $\cos \theta$, and the surface tension of the wetting liquid, γ_{LV}. Zisman defined the critical surface tension, γ_c, of the material under study by the value to which γ_{LV} extrapolates as $\cos \theta$ tends to unity, i.e. as θ tends to $0°$. Thus, γ_c is the surface tension of a liquid which will just spread on the surface of the material giving a zero contact angle. However, Zisman always emphasised that γ_c is *not* the surface free energy of the material but only an empirical parameter, the relative values of which act as one

TABLE 2
VALUES OF SURFACE FREE ENERGIES

Solid surface	γ_C (mN/m)	Surface free energy (mJ/m²)			Ref.
		γ_S^D	γ_S^P	γ_S	
Low-energy materials					
Polytetrafluoroethylene	18·5	14·5	1·0	15·5	28
Polyethylene	31	31·4	2·1	33·5	28
Polystyrene	33	38·4	2·2	40·6	28
Polychloroprene	38				29
Poly(methyl methacrylate)	39	35·9	4·3	40·2	13,29
Amine-cured epoxy	41	41·2	5·0	46·2	30
Rubber-modified epoxy		37·2	8·3	45·5	31
Phenol–resorcinol resin	52				29
Urea–formaldehyde resin	61				29
High-energy materials					
SiO_2	40[a]	78	209	287	32,33
Al_2O_3	45[a]	100	538	638	33,34
Fe_2O_3	46[a]	107	1250	1357	33–35

[a] For high-energy materials $\gamma_c \approx \gamma_{SV}$. The difference in γ_c and γ_S values reflects the adsorption of the test liquid vapour and atmospheric contaminants on to the high-energy surface. This is stated by

$$\gamma_S - \gamma_{SV} = \pi_S \tag{2}$$

For comparison, values for water (mJ/m²): $\gamma_{LV}^D = 22·0$; $\gamma_{LV}^P = 50·2$; $\gamma_{LV} = 72·2$.

would expect the surface free energy to behave. Representative values of γ_c are shown in Table 2.

An alternative approach has its roots in the proposal of Fowkes[10] that the surface free energy of a pure phase may be represented by the sum of contributions arising from the different types of force components. Fowkes[11] has identified at least seven components but Schultz *et al.*[12] have suggested that the surface free energy, γ, may be generally expressed by two terms, namely a dispersion, γ^D, and a polar, γ^P, component such that:

$$\gamma = \gamma^D + \gamma^P \tag{6}$$

Fowkes then proposed, following several well-established precedents, that the geometric mean of the dispersion force components is a reliable prediction of the interaction energies at the interface caused by dispersion forces; and Owens and Wendt[13] and Kaelbe and Uy[14] have proposed a similar geometric mean relation for the polar force interactions. These proposals may be employed to give a relation for the interfacial free energy for a solid–liquid system, namely:

$$\gamma_{SL} = \gamma_S + \gamma_{LV} - 2(\gamma_S{}^D \gamma_{LV}{}^D)^{\frac{1}{2}} - 2(\gamma_S{}^P \gamma_{LV}{}^P)^{\frac{1}{2}} \tag{7}$$

This relationship may be combined with eqn (3) to yield:

$$1 + \cos \theta = \frac{2(\gamma_S{}^D \gamma_{LV}{}^D)^{\frac{1}{2}}}{\gamma_{LV}} + \frac{2(\gamma_S{}^P \gamma_{LV}{}^P)^{\frac{1}{2}}}{\gamma_{LV}} \tag{8}$$

It has been assumed that the spreading pressure, π_S, for a liquid on a surface with which it makes finite contact is negligible (in effect this means that a relatively high-energy liquid is not adsorbed and hence cannot reduce γ_S of a low-energy solid). The values of γ_{LV}, $\gamma_{LV}{}^D$ and $\gamma_{LV}{}^P$ for many test liquids are known or can be readily determined; consequently if the contact angles of two liquids on a solid surface are measured then simultaneous equations may be formed from eqn (8) and solved to obtain values of $\gamma_S{}^D$ and $\gamma_S{}^P$. The total surface free energy, γ, of the solid surface is then simply the sum of these components. This approach has been discussed in detail by Kaelbe[15] and Sherriff.[16] Values of γ_S, $\gamma_S{}^D$ and $\gamma_S{}^P$ for various surfaces determined using the above technique are given in Table 2.

In the case of high-energy solid surfaces, a variety of techniques have been employed[17-25] to ascertain the surface free energy of such materials. Many involve the melting of the solid so the determination was performed at relatively high temperatures, but temperature coefficients of the surface free energy have been reported[17,23-25] to enable room-temperature values to be deduced. Other techniques[12,26,27] include measuring the contact angle of water on the solid under an *n*-alkane, or determining π_S from adsorption studies. Values of surface energies for some high-energy solids relevant to the present discussions are given in Table 2.

If we make the usually reasonable assumption that the surface free energies of the liquid and solid phases of a given material are not appreciably different then, from the data shown in Table 2 and eqns (4) and (5), the basic underlying reasons for several well-established aspects of adhesives technology may be discerned. For example, the low surface free energies of polytetrafluoroethylene and polyethylene results in non-wetting and hence poor interfacial contact being established when a typical liquid adhesive, e.g. an epoxy, is used to bond these materials. Therefore, when these materials are to be adhesively bonded they are usually subjected to a surface pretreatment which is designed to increase significantly their surface free energies[36,37]. On the other hand, the thermodynamics for the wetting of metal oxides and ceramics by organic adhesives are far more favourable, assuming their surfaces are clean in a technological sense, i.e. free from gross contamination which would act as a weak boundary layer and give weak joints. From previous comments, such basically high-energy adherend surfaces would be expected to have atmospheric contaminants adsorbed on to them but the polar nature of the resins which are used as structural adhesives will lead to displacement of the less polar, often hydrocarbon, atmospheric contaminants. Thus, a high degree of interfacial contact between the adhesive and the high-energy adherend surfaces may be attained and, furthermore, the high values of γ_S for the adherend will result in good intrinsic adhesion (see section 2.2.4, eqns (10) and (11)). These aspects are reflected by the relatively high joint strengths which can be readily achieved between, for example, an epoxy adhesive and metallic adherends. As discussed and illustrated later, the frequent use of a complex series of surface pretreatment stages for metallic adherends stems from the need to *maintain* the high joint strengths during the service-life of the bonded structure, and is *not* usually needed simply to achieve high initial strengths.

2.1.3. Kinetics of Wetting
Several workers[1,38,39] have stressed the importance of wetting as a kinetic process: although the thermodynamics may indicate the establishment of intimate molecular contact, the kinetics of wetting may be the determining factor.

The adhesive, at some time during the bonding operation, should possess a relatively low viscosity, e.g. no more than a few centipoises. However, the topography of the substrate surface may also influence the kinetics of wetting. The effect of surface topography is complex[1] but in most structural bonding applications the rate and extent of interfacial contact between adhesive and substrate will be increased the greater the rugosity of the substrate surface.

2.1.4. The Bonding Operation
The bonding operation should be conducted in an environment as free as possible of atmospheric contamination and possessing a low, or moderate, relative humidity. These factors will reduce to a minimum

the adsorption of atmospheric water, hydrocarbons, etc., on to freshly pretreated metallic adherends. In this respect the presence of fluorine- or silicone-containing contaminants, from mould-release agents for example, should particularly be avoided.

The bonding operation should also be conducted in a manner such as to prevent any air entrapment either at the interface or in the adhesive. Such air voids may be eliminated if the joint is bonded in any autoclave where there is a hydrostatic pressure high enough to compress entrapped air to a negligible volume, or by employing a 'vacuum-release' technique.[40] The latter method simply involves starting the cure in a vacuum (*ca* 650 Pa) and subsequently releasing the vacuum at the temperature at which the viscosity of the adhesive is at a minimum.

2.2. Mechanisms of Adhesion
In the case of a structural adhesive, once interfacial molecular contact between the adhesive and the adherends is obtained then the adhesive is cured, or hardened, so that it can transmit stresses from one adherend to the other. The question then arises, what is the nature of the intrinsic forces which act across the interfaces to prevent them separating under an applied load? The mechanisms of adhesion are still not fully understood and many theories are to be found in the current literature. The four main mechanisms of adhesion which have been proposed are:

(a) mechanical interlocking,
(b) diffusion theory,
(c) electronic theory,
(d) adsorption theory.

2.2.1. Mechanical Interlocking
This theory proposes that mechanical keying, or interlocking, of the adhesive into the irregularities of the substrate surface is the major source of intrinsic adhesion. However, the attainment of good adhesion between smooth surfaces exposes this theory as not being of general applicability. Indeed, there are only a few instances[1] where this mechanism has been conclusively demonstrated to play a dominant role and the enhancement of joint strength that may sometimes result from increasing the rugosity of the adherend surface appears often to result from other factors, e.g. an increase in surface area, improved

kinetics of wetting (see above) or an increase in the extent of plastic deformation of the adhesive.[41–45]

2.2.2. Diffusion Theory

Voyutskii[46] is the chief advocate of the diffusion theory of adhesion which states that the intrinsic adhesion of high polymers to themselves (autohesion), and to each other, is due to mutual diffusion of polymer molecules across the interface. This requires that the macromolecules or chain segments of the polymers (adhesive and substrate) possess sufficient mobility and are mutually soluble, i.e. they possess similar values of the solubility parameter. The solubility parameter, δ, may be defined by:

$$\delta = \left(\frac{\Delta H_v - \mathbf{R}T}{V}\right)^{\frac{1}{2}} \qquad (9)$$

where ΔH_v is the molar heat of vaporisation, $\mathbf{R}$ is the gas constant, T is the temperature (K) and V is the molar volume.

The above conditions are usually met in the autohesion of elastomers and in the solvent welding of compatible, amorphous plastics. In both these examples interdiffusion does significantly contribute to the intrinsic adhesion. However, where the solubility parameters of the materials are not similar, or where one polymer is highly crosslinked, is crystalline or is above its glass transition temperature, then interdiffusion is an unlikely mechanism.

2.2.3. Electronic Theory

If the adhesive and substrate have different electronic band structures there is likely to be some electron transfer on contact to balance Fermi levels which will result in the formation of a double layer of electrical charge at the interface. The electronic theory of adhesion is due primarily to Deryaguin and co-workers,[47–49] who have suggested that the electrostatic forces arising from such contact or junction potentials may contribute significantly to the intrinsic adhesion. The controversy this theory has caused is due to this final statement that electrostatic forces are an important *cause*, rather than merely a *result*, of high joint strength. Deryaguin's theory essentially treats the adhesive/substrate system as a capacitor which is charged due to the contact of the two different materials. Separation of the parts of the capacitor, as during interface rupture, leads to a separation of charge and to a potential difference which increases until a discharge occurs. Adhesion is presumed to be due to the existence of these attractive forces across the electrical double layer. This theory requires a variation of the measured work of adhesion with the pressure of the gas in which the adhesive measurements are conducted, but Weidner[50] has reported no increase in the peel strength of pressure-sensitive tapes when tested in vacuum as opposed to atmospheric pressure. Further, Deryaguin has equated the measured work of adhesion, determined from peel tests, with the calculated electrical energy stored in a capacitor and reported good agreement between the two quantities. However, the majority of the measured work of adhesion is dissipated through viscous and visco-elastic responses of the materials and this energy should not have been included in the value equated with the electrical energy.

Thus, apart from a few special circumstances[1] which are beyond the scope of the present text, the current view is that interfacial forces arising from any electrical double layer caused by different Fermi levels of the adhesive and adherends do not make a major contribution to the intrinsic adhesion.

2.2.4. Adsorption Theory
The adsorption theory of adhesion is the most generally accepted and has been discussed in depth by Kemball,[51] Huntsberger,[38] Staverman[52] and Wake.[53] This theory proposes that, provided sufficiently intimate intermolecular contact is achieved at the interface, the materials will adhere because of the surface forces acting between the atoms in the two surfaces; the most common such forces are van der Waals forces and are referred to as secondary bonds. In addition, chemisorption

TABLE 3
BOND TYPES AND TYPICAL BOND ENERGIES[54,55]

Type	Bond energy $(kJ\,mol^{-1})$
Ionic	590–1050
Covalent	63–710
Metallic	113–347
Permanent dipole–dipole interactions	
Hydrogen bonds involving fluorine	$\leqslant 42$
Hydrogen bonds excluding fluorine	10–26
Other dipole–dipole (excluding	
hydrogen bonds)	4–21
Dipole-induced dipole	<2
Dispersion (London) forces	0·08–42

may well occur and thus ionic, covalent and metallic bonds may operate across the interface; these types of bonds are referred to as primary bonds. The terms primary and secondary are in a sense a measure of the relative strength of the bonds as may be seen from Table 3, where the types of bonds are shown with estimates of the range of magnitude of their respective bond energies.

Secondary force interactions. Huntsberger[56] and others[57,58] have calculated the attractive forces between two planar bulk phases due solely to dispersion forces and have shown, for example, that even at a separation of 1 nm the attractive force would be approximately 100 MPa. This is considerably higher than the experimental strength of most joints. This discrepancy between theoretical and experimental joint strengths is attributed to air voids, cracks, defects or geometric features acting as stress raisers when the joint is loaded, causing rupture of the joint at stresses very much below the theoretical value. However, this calculation does indicate that high joint strengths may result from the intrinsic adhesion arising solely from dispersion force interactions.

The surface free energies, which may be determined as described in section 2.1.2, may be used to determine the intrinsic work of adhesion arising from interfacial secondary force interactions. The thermodynamic work of adhesion, W_A, required to separate a unit area of two phases forming an interface may be related to the surface free energies by the Dupré equation. In the absence of chemisorption, mechanical interlocking and interdiffusion the reversible work of adhesion, W_A, in an inert medium may be expressed by:

$$W_A = \gamma_S + \gamma_{Ad} - \gamma_{SAd} \qquad (10)$$

where γ_S represents the substrate, or adherend, and γ_{Ad} the now-cured adhesive. Using eqn (7) to eliminate the interfacial free energy, γ_{SAd}, then

$$W_A = 2(\gamma_S^D \gamma_{Ad}^D)^{\frac{1}{2}} + 2(\gamma_S^P \gamma_{Ad}^P)^{\frac{1}{2}} \qquad (11)$$

As may be seen from Table 3, the strength of the hydrogen bond lies between those of the van der Waals forces and the primary bonds, and the formation of hydrogen bonds across the interface may certainly increase the intrinsic adhesion as has been discussed by Pritchard.[59] He cites the dipping of nylon cords into a complex adhesive mixture of rubber and resorcinol–formaldehyde, as used in the production of tyres, as an example where hydrogen bonding may play an important role; the resorcinol–formaldehyde is adsorbed on to the nylon surface via hydrogen bonds through the phenolic groups.

It has also been suggested[60,61] that acid–base interactions across an interface may contribute to intrinsic adhesion forces. The analysis presented by Bolger and Michaels[61] attempts to identify many of the distinguishing features of acid–base interactions in terms of the ionic character of a hydroxylated metal oxide substrate and the organic functional group provided by the adhesive.

Primary interfacial bonding. Although it is evident that intrinsic adhesion arising from secondary bonding forces alone may result in adequate and high joint strengths, many adhesion scientists, including the author, believe that the additional presence of primary bonding may often increase the measured joint strength and is certainly often of great benefit in securing environmentally stable interfaces. However, although there is considerable indirect evidence[62-74] emphasising the importance of interfacial chemical bonding, studies which directly confirm its role are scarce. The use of sophisticated, surface-specific, analytical techniques are discussed in detail in Chapter 2, but methods such as laser–Raman spectroscopy,[75] X-ray photoelectron spectroscopy,[76] secondary-ion mass spectroscopy[77,78] and inelastic electron tunnelling spectroscopy[79] have all produced definitive evidence that primary interfacial bonding may occur in certain circumstances and may make a significant, indeed often vital, contribution to the intrinsic adhesion.

2.3. Concluding Remarks

As was mentioned earlier, structural adhesive joints when initially made and tested in an unaged state rarely exhibit failure at, or near, the interface. Thus, although the concepts above are obviously essential in order to understand and to be able to attain adequate adhesion, they rarely enter into a discussion on the mechanical properties of joints tested under these circumstances. However, any consideration of the durability of structural adhesive joints inevitably invokes surface and interfacial aspects: this is evident in the discussion of environmental attack in the following section.

3. ENVIRONMENTAL FAILURE

In this broad overview, some general observations are made on the durability of structural adhesive joints before some of the main aspects involved in the environmental failure of such joints are considered.

3.1. General Observations
It is of interest to distil from the numerous outdoor and accelerated laboratory trials reported, many of which are extensively discussed in later chapters, some general observations concerning the durability of structural adhesives.

3.1.1. Mechanical Performance and the Service Environment
In essence, it has been found that the mechanical properties of a bonded component may rapidly deteriorate upon exposure of the joint to its normal operating environment. Indeed, in some instances the loss of load-bearing capability has been so marked that premature joint failure has occurred at very low load levels and within a small fraction of the required service life. Empirical laboratory investigations established[80-82] many years ago that water, either in the liquid or vapour form, is the most hostile environment for structural adhesive

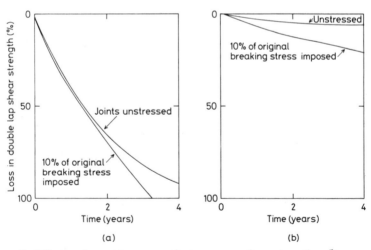

Fig. 3. *Effect of outdoor weathering on the strength of epoxy-polyamide/aluminium-alloy (chromic-acid etch pretreatment) structural joints. (a) Hot/wet tropical site; (b) hot/dry desert site.*

joints that is commonly encountered. For example, in natural outdoor climatic trials it is invariably the presence of moisture which is responsible for environmental attack upon the adhesive or the interfacial regions rather than, say, oxygen or ozone. Further, the rate of environmental attack is greater if the temperature is relatively high or the joint is subjected to stress. These aspects are illustrated[83] in Fig. 3, which imparts an appreciation of the extent of the problem. It is evident that the hot/wet tropical environment was by far the more hostile and that the presence of an applied stress (of about 3·7 kN) increased the rate of loss of strength. Indeed, the joints exposed under this stress level at the tropical site fell apart shortly after three years had elapsed.

3.1.2. Locus of Joint Failure

Whilst the locus of failure of unaged structural joints is invariably by cohesive fracture in the adhesive layer, after environmental attack it is usually by apparent interfacial failure between the adhesive (or primer) and the adherend. However, whether the failure path is truly at the interface, or whether it is in the oxide, in the boundary layer of the adhesive or in the primer layer (if present), is a matter of some debate and controversy, as is apparent from the discussions in later chapters. The exact failure path probably depends upon the particular joint under examination but, wherever it is, the above observation does largely concentrate our attention on the interfacial regions if we are to understand and prevent environmental failure of structural adhesive joints.

3.1.3. Parameters Affecting Environmental Resistance

The following parameters have been found to affect joint durability:

Environment. The presence of an hostile environment is the essence of the problem and water has been observed to be the most harmful and most commonly encountered environment.

Concentration of water. This is discussed in detail in Chapter 3, but essentially the higher the activity of the water present in the environment, the more rapid and greater the degree of attack. Furthermore, there often appears to be a minimum concentration below which no environmental failure, or at least no significant attack over a comparatively long time-scale, occurs.

Temperature. Increasing the temperature of the environment increases the rate of strength loss.

Adhesive type. The chemical type of structural adhesive employed can affect both the rate and degree of environmental failure and this may arise for several reasons. First, obviously, if the service environment physically or chemically attacks the adhesive to any significant extent, the joint may well be appreciably weakened. However, the preceding comments reveal that, whilst structural adhesives may suffer from reduction of modulus and some loss of strength due to plasticisation by water especially when operating near their glass transition temperature, it is the interfacial regions which represent the critical area. Second, however, the adhesive type does appear to influence the environmental stability of the interfacial regions. For example, especially noteworthy is the often-superior joint durability shown by the older phenolic-based structural adhesives, particularly those containing a nitrile rubber to improve their toughness, compared to the more modern epoxy-based adhesives. Nevertheless, the epoxy types are now generally preferred on the basis of their lower cure-temperature/pressure requirements and usually superior toughness and, hence, peel strengths. These aspects are discussed in detail in Chapters 3 and 4.

Adherend. Joints to metallic adherends present the main problem. Joints to plastics, grp and cfrp are far less susceptible to environmental attack; that is not to say that such joints never suffer from such attack, but when this does occur it is usually the adherend, e.g. the composite material, which is itself attacked more readily and rapidly than the bulk adhesive or the interfacial regions. Metallic adherends are discussed in detail in Chapters 5, 6 and 7 and composite adherends are briefly mentioned in Chapter 3. The long practical experience of the aerospace industries with many of these materials is discussed in Chapter 8.

Adherend surface pretreatment. This is an extremely important factor. To ensure initially strong joints it is usually sufficient to remove surface contamination, weak oxide layers, etc., which may act as weak boundary layers. However, to produce durable joints it is also necessary to form stable oxides which are 'receptive' to the adhesive and, also, frequently to employ a specially developed primer. There is broad agreement with respect to the pretreatments which impart the best

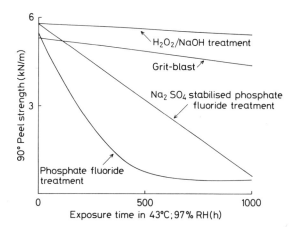

Fig. 4. Peel strength of titanium-alloy/epoxy joints as a function of surface pretreatment of titanium adherends (titanium (6 Al 4 V) alloy). Adhesive: 120°C curing epoxy. (Note that the various pretreatments have a relatively small effect upon the initial, unaged joint strength.)

durability but there is still much speculation concerning how these pretreatments actually lead to good environmental resistance. New surface analytical techniques have greatly assisted the scientific interpretation of the mechanisms involved, as reviewed in Chapters 2 and 3. As an example of the effect of adherend surface pretreatment upon joint durability the data[83] shown in Fig. 4 clearly illustrate the importance of selecting an adequate pretreatment for attaining the required service-life. However, it should also be noted that the effectiveness of a given pretreatment is sometimes dependent upon the actual alloy type and manufacturing process employed.

Applied stress. As illustrated in Fig. 3, the rate of loss of strength will be faster if a tensile or shear stress is present, albeit an externally applied stress or internal stresses induced by adhesive shrinkage (incurred during cure) or by adhesive swelling (due to water uptake). Such stresses render primary and secondary bonds more susceptible to environmental attack by lowering the free energy barrier that must be crossed if the bond is to change from an unbroken to a broken state, i.e. they lower the activation energy for, and so increase the rate of, bond rupture. Stress also probably increases the rate of diffusion of the ingressing medium. On the positive side, plasticisation of the adhesive

A. J. Kinloch

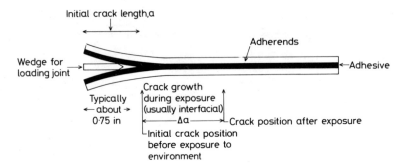

Fig. 5. *The Boeing wedge-test for assessing joint durability. Typical exposure (to hot/wet environment) is 1 h, at 49°C, 100% R.H. Environmental crack growth, Δa, is then determined, which is a measure of joint durability.*

may diminish stresses by stress relaxation and crack blunting mechanisms. Indeed, crack-tip blunting may actually cause the apparent toughness of the adhesive to increase, and such an effect has been reported after short exposure times insufficient for significant interfacial attack.

Joint design. Since it is the interfacial regions in which environmental failure initiates, a joint design which has a relatively high stress concentration at, or near, the interface will tend to reveal durability effects more readily. The standard peel test (Fig. 2) or the more recently developed Boeing wedge-test (Fig. 5) have greater sensitivities towards environmental attack than, say, the conventional lap joint. Indeed, the Boeing wedge-test has been found to be particularly useful as a quick and convenient quality control technique for assessing the adequacy of the adherend surface pretreatment with respect to joint durability; a value of environmental crack growth, Δa, is specified which must not be exceeded, so giving a pass/fail criterion.

3.2. The Interfacial Regions

3.2.1. Interface Stability
Thermodynamic considerations. The intrinsic stability of the adhesive/adherend interface in the presence of any liquid environment may be assessed from the thermodynamic arguments advanced by Gledhill and Kinloch.[30]

As defined in section 2.2.4., the thermodynamic work of adhesion, W_A, is the energy required to separate unit area of two phases forming an interface. If only secondary (e.g. van der Waals) forces are acting across the interface then the value of W_A in an inert medium may be expressed by eqn (10). However, in the presence of a liquid (denoted by the suffix L) the work of adhesion, W_{AL}, is now given by:

$$W_{AL} = \gamma_{SL} + \gamma_{AdL} - \gamma_{SAd} \qquad (12)$$

where γ_{SL} and γ_{AdL} are the interfacial free energies between the adherend/liquid and adhesive/liquid interfaces respectively.

For an adhesive/adherend interface in an inert environment the work of adhesion, W_A, has a positive value, indicating a thermodynamic stability of the interface. However, in the presence of a liquid the value of W_{AL} may well have a negative value indicating the interface is now unstable and will dissociate. Thus calculation of the terms W_A and W_{AL} enables the environmental study of the interface to be predicted. Some examples of W_A and W_{AL} are shown in Table 4 and the generality of this approach is illustrated by reference to other environments besides water. For those interfaces where there is a change from a positive to a negative work of adhesion, this provides a driving force for the displacement of adhesive on the adherend surface by the liquid. It is therefore to be expected that if the joint is subjected to such an environment there will be a progressive encroachment into the joint of debonded interface. This will have the effect of progressively changing the locus of joint failure to interfacial between adhesive and adherend. The thermodynamic approach also reveals that, since both metal oxides and water are relatively polar (see Table 2), water will have a tendency to be preferentially adsorbed on to an oxide surface and so create a weak boundary layer between the adhesive (or primer) and metallic adherend. Thus, high-energy, polar adherends, will be the most difficult materials to bond adhesively with respect to ensuring a long service-life in the presence of moisture, an observation in accord with industrial experience. Also of interest is that the epoxy adhesive/cfrp interface is predicted to be relatively stable, again in accord with experimental observation.

In those instances where W_{AL} is not negative but $W_A > W_{AL} \geqslant 0$, the input of additional work is a necessary requisite for joint failure. However, as might be expected, the measured joint strength for interfacial failure is now reduced by the presence of the liquid.

Finally, it should be noted that the thermodynamics as outlined

A. J. Kinloch

TABLE 4
VALUES OF W_A AND W_{AL} FOR VARIOUS INTERFACES AND ENVIRONMENTS

Interface	W_A in inert medium	Work of adhesion (mJ/m^2) In liquid environment		Interfacial debonding[a]	Ref.
		Liquid	W_{AL}		
Epoxy adhesive/ferric oxide (mild steel)	291	Ethanol	22	No	30
		Formamide	−166	Yes	30
		Water	−255	Yes	30
Epoxy adhesive/aluminium oxide	232	Water	−137	Yes	33
Epoxy adhesive/silica	178	Water	−57	Yes	33
Epoxy adhesive/carbon-fibre-reinforced plastic	88 → 90	Water	22 → 44	No	87
Vinylidene chloride–methylacrylate copolymer/polypropylene	88	Water	37	No	88
		Sodium n-octylsulphate soln	1·4	No	88
		Sodium n-dodecylsulphate soln	−0·9	Yes	88
		Sodium n-hexadecylsulphate soln	−0·8	Yes	88

[a] Upon immersion of unstressed joints was interfacial debonding observed?

above, and discussed in more detail in Chapters 3 and 4, take no account of interfacial adhesion forces arising from primary bonds or mechanical interlocking. Further, they provide no information on the kinetics of the failure mechanism and hence no *quantitative* prediction as to the expected service-life, for which the thermodynamic analysis needs to be combined with either (i) a stress-biased activated rate theory, as developed by Zhurkov and co-workers[84] and used in joint fracture studies by Levi *et al.*,[85] or (ii) a continuum fracture mechanics approach, as developed by Kinloch *et al.*[86] and discussed in Chapter 3.

Interfacial secondary forces. The relations outlined above may be used to predict, for a given adherend and liquid environment, the required values of γ_{AD}^D and γ_{Ad}^P of the adhesive if adhesive/adherend interface stability is to be maintained in the presence of the environment with only secondary forces acting across the interface. Hence, the surface chemical properties of the adhesive needed for interface stability may be calculated and then employed in the adhesive selection process.

This concept is illustrated in Fig. 6 for a mild steel (ferric oxide) adherend, with ethanol as the potential liquid environment in Fig. 6(a) and water in Fig. 6(b). Values of γ_{Ad}^D and γ_{Ad}^P of possible adhesives can be measured, although the values of many common adhesives may be found in the literature, as shown earlier in Table 2. For example, in Fig. 6 the value of γ_{Ad}^D and γ_{Ad}^P for an epoxy-amine structural adhesive and a fluorinated rubbery adhesive are given. Thus, examination of Fig. 6(a) reveals that in selecting an adhesive for bonding mild steel with ethanol a likely service environment, employing the epoxy-amine adhesive would result in an environmentally stable interface whilst using the fluorinated rubbery adhesive would not. However, for bonding mild steel, with an aqueous environment likely to be encountered by the joint, Fig. 6(b) reveals that neither adhesive would provide an intrinsically stable interface. Indeed, the data confirm that if only secondary forces are acting across the interface water will virtually always desorb organic adhesives, which typically have low surface free energies of less than *ca* 60 mJ/m^2, from a metal oxide surface. Hence, for such interfaces, stronger forces must be forged which are more resistant to rupture by water.

Interfacial primary forces. An approximate indication of the contribution to interface stability from the additional establishment of

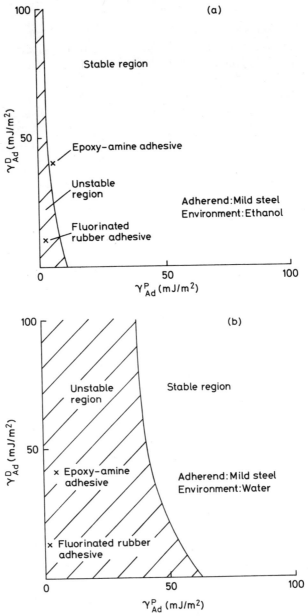

Fig. 6. *Predictions of interface stability for mild-steel adherends as a function of dispersion, γ_{Ad}^{D}, and polar, γ_{Ad}^{P}, force components of the surface free energy of possible adhesives. (a) In ethanol environment; (b) in aqueous environment.*

primary, chemical, bonds across the interface may be deduced by taking the chemical bond energy to be of the order of 250 kJ/mol (see Table 3) and assuming a coverage of 25 Å^2 per adsorbed site. This yields an intrinsic work of adhesion of $+1650$ mJ/m^2 and from energetic considerations it would be unlikely that water would readily displace such a chemisorbed layer.

In the structural bonding of metals the establishment of interfacial primary bonds has largely been achieved by the use of phenolic[61,79] or silane-based primers,[77,78,79] as discussed in detail in Chapter 3. Gettings and Kinloch[77] have investigated a range of silane-based primers deposited on a mild-steel substrate which resulted in adhesive joints possessing different durabilities. They also employed secondary ion mass spectroscopy (SIMS) and X-ray photoelectron spectroscopy (XPS) to identify the nature of the silane/metal oxide interfaces; the former technique proved especially rewarding. These techniques are reviewed in greater detail in Chapter 2. Basically, in SIMS ionised particles are ejected from the surface by the action of an argon beam and are mass-analysed. As the current densities used are low ($\sim 10^{-10}$ A/cm^2) the first one or two monolayers of the surface can be investigated. Either atoms or molecules can be ionised and thus details about the chemical state of atoms in a surface can be inferred. The SIMS spectrum from a silane primer (γ-glycidoxypropyltrimethoxysilane) coated on a mild-steel substrate is shown in Fig. 7. This primer resulted in considerable increases in joint durability compared to the unprimed adherends. The spectrum reveals the presence of SiO$_2$H$^-$, SiO$_2^-$ and FeSiO$^+$ radicals, none of which were detected on other silane-coated substrates where there was no improvement in joint durability. The SiO$_2$H$^-$ and SiO$_2^-$ radicals indicate that only with the good primer did polymerisation occur to give a polysiloxane structure on the metal substrate—these radicals arose from the polysiloxane. The FeSiO$^+$ radical provides strong evidence for the formation of a chemical bond, probably —Fe—O—Si≡, between the metal oxide and the polysiloxane primer. Thus only for the silane primer where there was evidence of chemical, rather than purely secondary, bonding between the primer and the metal oxide was there observed any improvement in joint durability. Similar studies[78] have been conducted on stainless steels and again only those interfaces where there was evidence of chemical bonding (—Fe—O—Si≡ *and* —Cr—O—Si≡) were there observed significant increases in environmental resistance of joints. These results suggest that the reaction

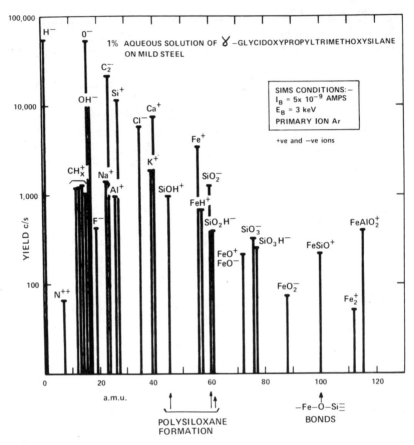

Fig. 7. *Positive and negative SIMS spectrum from a mild-steel substrate coated with γ-glycidoxypropyltrimethoxysilane.*

mechanism of the silane primer which is responsible for the increase in the environmental resistance of the joint is probably as shown in Fig. 8.

Finally, work using Auger and X-ray photoelectron spectroscopy has shown that, although a silane primer may increase interface stability and so enhance the service performance of the bonded component, the polysiloxane is then itself the weakest part of the joint and environmental failure occurs by cohesive fracture of this layer[89] (which is only a few tens, or so, of nanometres thick). The mechanism of attack

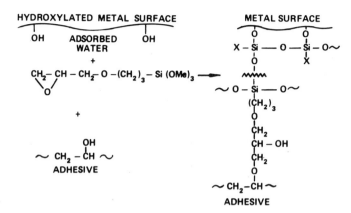

Fig. 8. *Proposed mechanism of adhesion for epoxy adhesive/γ-glycidoxy-propyltrimethoxysilane/ferric oxide interfaces.*

probably involves hydrolysis of the —Si—O—Si— bonds so, to improve further the service performance, attention needs to be focused on increasing the intrinsic resistance of the silane-based primers.

Mechanical interlocking. Results[90,91] in the literature clearly demonstrate that those surface pretreatments which produce the greatest degree of macro-roughness on the adherend surface, as a grit-blasting pretreatment invariably does, are not necessarily those which impart the best environmental resistance to the subsequent joint. Nevertheless, mechanical interlocking, but on a micro-scale, has recently been suggested to be a major mechanism in increasing interface stability in certain epoxy/aluminium alloy joints when a phosphoric acid anodising surface pretreatment is employed for the aluminium alloy prior to adhesive bonding.

Phosphoric acid anodising is a particularly effective surface treatment for aluminium alloys for ensuring very good service performance of joints in aqueous environments. This pretreatment has been shown to produce an open, deep porous oxide layer on the aluminium, as illustrated in Chapters 2 and 5. The adhesive, or primer, appears to penetrate quite deeply into the porous oxide and it has therefore been suggested that this produces micro-interlocking which imparts good interface stability. However, since this pretreatment also results in an environmentally stable oxide layer, unlike many other common pretreatments, it is difficult to identify the detailed mechanism whereby it imparts good service performance. Contributions may arise from a micro-mechanical interlocking mechanism, increasing interface stability, or increased oxide stability, or even the formation of primary interfacial bonds which have not as yet been identified. However, in the surface pretreatment of aluminium alloys one mechanism, namely oxide stability, has recently been shown to be of major importance and is briefly introduced in the next section.

3.2.2. Oxide Stability

Gross adherend corrosion. Gross adherend corrosion is *not* usually a major mechanism of environmental failure. For example, corrosion of the surface of a metallic adherend is often a post-failure phenomenon, occurring after the displacement of adhesive on the metal oxide by water.[30] Only in special circumstances, e.g. with clad aluminium alloys or in a salt-water environment, is gross corrosion of the adherend an important failure mechanism.

The potential problem with clad aluminium alloy is of particular interest and has been studied in detail by Riel.[92] With clad aluminium alloys the electrode potential of the cladding is generally higher than that of the base alloy. This choice is deliberate: the clad material is selected to be anodic with respect to the base alloy so that in a corrosive environment the cladding will be consumed preferentially, thus protecting the base alloy. The mechanism is very effective in protecting the structure from surface corrosion such as pitting, which is less likely to occur on clad alloy due to the nature of the alloy; where pits do form and penetrate the clad surface its anodic nature will cause the pit to grow laterally once the base alloy is reached, rather than penetrating it as with unclad materials. However, whilst this mechanism of corrosion protection inhibition may be effective for exposed aluminium-alloy structures, if one considers the mechanisms whereby clad aluminium alloy achieves its corrosion resistance, the clad layer may be actually undesirable in the context of adhesive bonding. A galvanic cell may be established between cladding and base alloy with the progressive destruction of the interfacial regions. In the United States aerospace industry the current trend is away from adhesive bonding to clad aluminium alloys,[83,93,94] but where unclad alloys are bonded and used in areas exposed to corrosive environments any non-bonded, exterior surfaces must be protected by appropriate means in order to limit surface corrosion.

Oxide transformation. As mentioned previously, the choice of surface pretreatment for the adherend prior to adhesive bonding has a major effect on the subsequent service performance of the joint in hostile environments. This is particularly marked in the case of aluminium alloys and some typical results are shown in Fig. 9, although it should be noted that the exact ranking order may change slightly depending upon the adhesive primer and type of alloy being employed.

There is considerable evidence that the importance of the surface pretreatment arises from the influence it has upon the surface chemistry and structure of the aluminium oxide generated. Hence, the choice of surface pretreatment may greatly affect the stability of the oxide upon exposure to water.

Early evidence for the initially-generated oxide changing its structure upon exposure to moisture comes from the work of Noland.[95] He reported that the oxide produced on aluminium alloys by a chromic–sulphuric acid etch, a common pretreatment in the aerospace industry, is unstable in the presence of moisture and postulated that the oxide changes to a weaker, gelatinous type which is hydrated and is termed 'gelatinous–boehmite'. Support for the transformation came from XPS analysis of the oxide surface before and after ageing and Fig. 10(a)

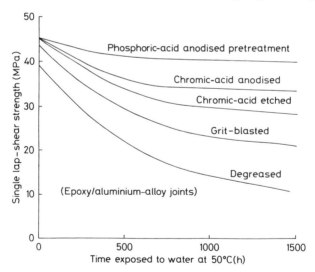

Fig. 9. *Effect of surface pretreatment on the performance of aluminium-alloy epoxy joints subjected to accelerated ageing in water at 50°C.*

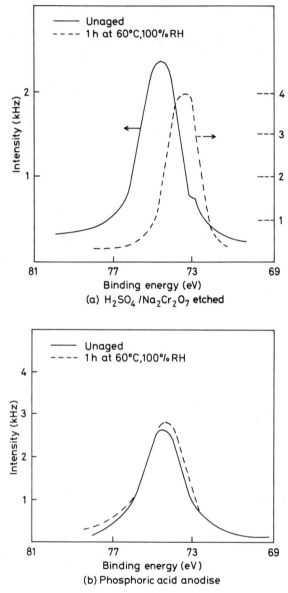

Fig. 10. *X-Ray photoelectron analysis of aluminium-alloy surface (Al 2p peak) before and after ageing. (a) Chromic–sulphuric acid-etch pretreatment; (b) phosphoric-acid anodised pretreatment.*

shows that a change in binding energy is observed for the aluminium $2p$ peak position, indicating a change in oxide structure. Noland also examined epoxy/aluminium alloy joints after exposure to hot, humid conditions and reported that, although from a visual inspection apparent interfacial failure had occurred, in fact the locus of failure was in the mechanically weak, gelatinous-boehmite oxide layer. Recent work (see Chapters 2 and 5) has employed electron diffraction and scanning transmission electron microscopy and essentially confirmed this earlier study. The original oxide formed by the pretreatment was found to be largely amorphous but upon exposure to moisture it became hydrated and changed to crystalline, pseudo-boehmite (i.e. a material containing somewhat more water than perfectly crystalline boehmite). This hydrated oxide could be readily distinguished by its distinctive morphology which consisted of irregularly shaped platelets, which the authors dubbed a 'cornflake structure'. This structure was however only loosely bound to the underlying oxide and thus represented a weak boundary layer, but one which was actually formed *in situ* in the joint. In direct contrast, Noland[95] found that the oxide produced by the phosphoric acid anodising pretreatment was basically stable in the presence of moisture, as may be seen from the XPS analysis in Fig. 10(b).

The detailed reasons why different pretreatments produce different oxides, which apparently possess very different stabilities in aqueous environments, have yet to be resolved. Nevertheless, some possible reasons have recently begun to emerge. Firstly, the phosphoric acid pretreatment leaves an oxide surface containing phosphorus, present as PO_4^{3-} ions, which may inhibit hydration of the oxide.[91,96,97,98] Secondly, some pretreatments with certain alloys result in the accumulation of elements such as copper[99] and magnesium[91,99,100] in the oxide layer and these elements sometimes appear to decrease the stability of the oxide. Indeed, Kinloch and Smart, using XPS, examined a range of pretreatments and found a correlation between high magnesium content in the oxide layer and poor joint durability, as indicated in Table 5. However, this and subsequent work[91,99,100] clearly confirmed that a low level of magnesium is not the sole criterion for attaining good durability, but that good durability may be unattainable if a high level of magnesium is present.

3.2.3. Stability of the Adhesive Boundary Layer

Early work[81] on the mechanisms of environmental failure concluded that hydrolysis in a boundary layer of the adhesive near the

TABLE 5

RELATION BETWEEN MAGNESIUM CONCENTRATION IN OXIDE OF ALUMINIUM
ALLOY AND SUBSEQUENT JOINT DURABILITY

Surface pretreatment for aluminium alloy	Al:Mg ratio in oxide surface[a]	Joint durability
Degreased	6:1	Very poor
Grit-blasted	15:1	Moderate
Chromic acid etch	57:1	Good
Phosphoric acid anodise	110:1	Very good

[a] Bulk Al:Mg ratio was 53:1.

adhesive/adherend interface was a primary failure mechanism in the epoxy/aluminium-alloy joints investigated. This mechanism has been largely rejected by subsequent investigators but recent studies by Brockmann and colleagues have re-focused attention on this possible failure mode, as discussed in Chapter 7. When considering this failure mechanism, it should be borne in mind that there is firm evidence that the boundary layer of adhesive adjacent to the metal oxide surface may well possess a different chemical and physical structure, e.g. lower crosslink density[101] or lower concentration of filler particles,[102] compared to the bulk adhesive and these differences may depend upon the activity of the adherend surface resulting from the particular pretreatment employed. Thus, the secondary adhesive layer could be more susceptible to hydrolysis than the bulk adhesive.

3.3. Kinetics of Failure

Several workers have shown that the kinetics of environmental failure may be largely governed by the rate of diffusion of water into the joint. Fortunately, water uptake by structural adhesives often behaves according to Fick's law so, from measuring the diffusion constant using bulk adhesive film samples, the water concentration profile within the joint as a function of geometry, temperature and water activity may be predicted. The concentration of water in a joint at any given time increases with the values of the latter two parameters, so some of the observations on the effect of temperature and water activity upon the durability of structural adhesive joints may be readily explained (Section 3.1.3). The diffusion of water into adhesives is described in

Chapter 3, where the relationships between water diffusion and loss of joint strength are investigated.

3.4. Concluding Remarks

The main aspects of the durability of structural adhesive joints, which will be explored in the following chapters, are summarised below

(a) Water is a particularly aggressive environment for adhesive joints and the service performance of structural adhesive joints may be seriously impaired by exposure to hot/wet environments. The loss of strength is accelerated by the presence of a tensile or shear stress.

(b) Adherends which possess high surface free energies, e.g. metals, lead to joints which are particularly susceptible to environmental failure, and attack usually occurs in the interfacial regions of the joint.

(c) The mechanisms of environmental failure which have been identified are:

(i) Displacement of adhesive on the adherend by water due to the rupture of secondary bonds at the adhesive/adherend interface. This may be predicted from thermodynamic considerations.

(ii) Subtle changes occurring in the oxide structure on the metallic adherend, e.g. hydration, which cause a mechanical weakening and, eventually, failure of the oxide layer. This process may be inhibited or accelerated by the presence of trace elements in the oxide layer.

(iii) Hydrolysis in a boundary layer of the adhesive, adjacent to the adherend surface, the properties of this boundary layer being different from those of the bulk adhesive.

(iv) In special circumstances only, for example with clad aluminium alloys or in a sea-water environment, is gross corrosion of the adherends an important mechanism. Usually corrosion is a post-failure phenomenon.

(d) The kinetics of environmental failure are influenced by the diffusion of water through the adhesive and in some instances this may be the rate-determining stage.

(e) To increase service performance, a surface pretreatment and/or a surface primer must be employed to ensure the generation of stable oxide layers and the formation of strong interfacial forces which are resistant to rupture by water.

REFERENCES

620·11 1. Kinloch, A. J., *J. Mater. Sci.*, **15** (1980), 2141.
2. Kinloch, A. J., *J. Mater. Sci.*, **17** (1982), 617.
3. Adams, R. D., in *Developments in Adhesives—2*, ed. A. J. Kinloch, Applied Science Publishers, London (1981), p. 45.
4. Hart-Smith, L. J., in *Developments in Adhesives—2*, ed. A. J. Kinloch, Applied Science Publishers, London (1981), p. 1.
5. Wake, W. C., *Adhesion and the Formulation of Adhesives*, 2nd Edn, Applied Science Publishers, London (1982).
6. Young, T., *Trans. Roy. Soc.*, **95** (1805), 65.
7. Good, R. J., in *Aspects of Adhesion—7*, ed. D. J. Alner and K. W. Allen, Transcripta Books, London (1973), p. 182.
8. Bangham, D. H. and Razouk, R. I., *Trans. Faraday Soc.*, **33** (1937), 1459.
9. Zisman, W. A., in *Adv. Chem. Ser.* **43**, ed. R. F. Gould, Amer. Chem. Soc., Washington (1964), p. 1.
10. Fowkes, F. M., *J. Phys. Chem.*, **67** (1967), 2538.
11. Fowkes, F. M., *J. Colloid Interf. Sci.*, **28** (1968), 493.
12. Schultz, J., Tsutsumi, K. and Donnet, J. B., *J. Colloid Interf. Sci.*, **59** (1977), 277.
13. Owens, D. K. and Wendt, R. C., *J. Appl. Polym. Sci.*, **13** (1969), 1740.
14. Kaelbe, D. H. and Uy, K. C., *J. Adhesion*, **2** (1970), 50.
15. Kaelbe, D. H., *Physical Chemistry of Adhesion*, John Wiley/Interscience, New York (1971), p. 153.
16. Sherriff, M., *J. Adhesion*, **7** (1976), 257.
17. Bondi, A., *Chem. Revs. Amer. Chem. Soc.*, **52** (1953), 417.
18. Dunning, W. J., in *Adhesion*, ed. D. D. Eley, Oxford University Press, London (1961), p. 57.
19. Salmon, G. S., in *Adhesion and Adhesives*, vol. 1, ed. R. Houwink and G. Salmon, Elsevier, Amsterdam (1965), p. 35.
20. Rasmussen, J. J. and Nelson, R. P., *J. Amer. Ceram. Soc.*, **54** (1971), 398.
21. Valentine, T. M., *Mater. Sci. Eng.*, **30** (1977), 205.
22. Valentine, T. M., *Mater. Sci. Eng.*, **30** (1977), 211.
23. Kingery, W. D., *J. Amer. Ceram. Soc.*, **37** (1954), 42.
24. Schonhorn, H. H., *J. Phys. Chem.*, **71** (1967), 4578.
25. Rhee, S. K., *Mater. Sci. Eng.*, **16** (1974), 45.
26. Schultz, J., Tsutsumi, K. and Donnet, J. B., *J. Colloid Interf. Sci.*, **59** (1977), 272.
27. Fowkes, F. M., *Ind. Eng. Chem.*, **56** (1964), 40.
28. Wu, S., *J. Macromol. Sci. Revs Macromol. Chem.*, **C10** (1974), 1.
29. Shafin, E. G., in *Handbook of Adhesives*, ed. I. Skeist, Van Nostrand, New York (1977), p. 67.
30. Gledhill, R. A. and Kinloch, A. J., *J. Adhesion*, **6** (1974), 315.
31. Lewis, A. F. and Gounder, R. T. N., in *Treatise on Adhesion and Adhesives*, vol. 5, ed. R. L. Patrick, Marcel Dekker, New York (1981), p. 349.

32. Shafin, E. G. and Zisman, W. A., *J. Amer. Ceram. Soc.*, **50** (1967), 478.
33. Dukes, W. A., Gledhill, R. A. and Kinloch, A. J., in *Adhesion Science and Technology*, ed. L. H. Lee, Plenum Press, New York (1975), p. 597.
34. Bernett, M. K. and Zisman, W. A., *J. Colloid Interf. Sci.*, **28** (1968), 243.
35. Gledhill, R. A., Kinloch, A. J. and Shaw, S. J., *J. Adhesion*, **9** (1977), 81.
36. Dukes, W. A. and Kinloch, A. J., in *Developments in Adhesives—1*, ed. W. C. Wake, Applied Science Publishers, London (1977), p. 251.
37. Brewis, D. M. (ed.), *Surface Analysis and Pretreatment of Plastics and Metals*, Applied Science Publishers, London (1982).
38. Huntsberger, J. R., in *Treatise on Adhesion and Adhesives*, vol. 1., ed. R. L. Patrick, Marcel Dekker, New York (1967), p. 119.
39. Bascom, W. D. and Patrick, R. L., *Adhesives Age*, **17**(10) (1974), 25.
40. Bascom, W. D. and Cottingham, R. L., *J. Adhesion*, **4** (1972), 193.
41. Evans, J. R. and Packham, D. E., *J. Adhesion*, **10** (1979), 177.
42. Bascom, W. D., Timmons, C. O. and Jones, R. L., *J. Mater. Sci.*, **19** (1975), 1037.
43. Mulville, D. R. and Vaishnav, R., *J. Adhesion*, **7** (1975), 215.
44. Bair, H. E., Matsuoka, S., Vadimsky, R. G. and Wang, T. T., *J. Adhesion*, **3** (1971), 89.
45. Wang, T. T. and Vazirani, H. N., *J. Adhesion*, **4** (1972), 353.
46. Voyutskii, S. S., *Autohesion and Adhesion of High Polymers*, John Wiley/Interscience, New York (1963).
47. Deryaguin, B. V., *Research*, **8** (1955), 70.
48. Deryaguin, B. V., Krotova, N. A., Karassev, V. V., Kirillova, Y. M. and Aleinikova, I. N., *Proc. 2nd Internat. Congress on Surface Activity—III*, Butterworths, London (1957), p. 417.
49. Deryaguin, B. V. and Smilga, V. P., *Adhesion, Fundamentals and Practice*, McLaren, London (1969), p. 152.
50. Weidner, C. L., *Adhesives Age*, **6**(7) (1963), 30.
51. Kemball, C., in *Adhesion*, ed. D. D. Eley, Oxford University Press, London (1961), p. 19.
52. Staverman, A. J., in *Adhesion and Adhesives*, vol. 1, ed. R. Houwink and G. Salmon, Elsevier, Amsterdam (1965), p. 9.
53. Wake, W. C., *RIC Lecture Series* No. 4 (1966), p. 1.
54. Pauling, L., *The Nature of the Chemical Bond*, Cornell University Press, New York (1960).
55. Good, R. J., *Treatise on Adhesion and Adhesives*, vol. 1, ed. R. L. Patrick, Marcel Decker, New York (1967), p. 15.
56. Huntsberger, J. R., *Adhesives Age*, **13**(11) (1970), 43.
57. Orowan, E., *J. Franklin Inst.*, **290** (1970), 493.
58. Tabor, D., *Rep. Prog. Appl. Chem.*, **36** (1951), 621.
59. Pritchard, W. H., in *Aspects of Adhesion—6*, ed. D. J. Alner, University of London Press, London (1970), p. 11.
60. Fowkes, F. M., in *Recent Advances in Adhesion*, ed. L. H. Lee, Gordon and Breach, London (1973), p. 39.
61. Bolger, J. C. and Michaels, A. S., in *Interface Conversions for Polymer Coatings*, ed. P. Weiss and G. D. Cheever, Elsevier, New York (1968), p. 3.

eLollis, N. J., *Adhesives Age*, **11**(12) (1968), 21.
63. DeLollis, N. J., *Adhesives Age*, **12**(1) (1969), 25.
64. Rutzler, J. E., *Adhesives Age*, **2**(7) (1954), 28.
65. Vilenskii, A. I., Virlich, E. E. and Krotova, N. A., *Soviet Plastics*, **5** (1973), 68.
66. Koldunovich, G. E., Epshtein, V. G. and Chekanova, A. A., *Soviet Rubber Technol.*, **29** (1970), 22.
67. Trostyanskaya, E. B., Golovkin, G. S. and Komarov, G. V., *Soviet Rubber Technol.*, **25** (1966), 13.
68. Dean, R. B., *Official Digest*, **36** (1964), 664.
69. Ahagon, A. and Gent, A. N., *J. Polym. Sci. Polymer Phys.*, **13** (1975), 1285.
70. Andrews, E. H. and Kinloch, A. J., *Proc. Roy. Soc.*, **A332** (1973), 401.
71. Gent, A. N. and Hsu, E. C., *Macromolecules*, **7** (1974), 933.
72. Lerchenthal, C. H., Brennan, M. and Yits'Haq, N., *J. Polymer Sci. Polymer Chem.*, **12** (1975), 737.
73. Lerchenthal, C. H. and Brennan, M., *Polymer Eng. Sci.*, **16** (1976), 747.
74. Lerchenthal, C. H. and Brennan, M., *Polymer Eng. Sci.*, **16** (1976), 760.
75. Koenig, J. L. and Shih, P. T. K., *J. Colloid. Interf. Sci.*, **36** (1971), 247.
76. Bailey, R. and Castle, J., *J. Mater. Sci.*, **12** (1977), 2049.
77. Gettings, M. and Kinloch, A. J., *J. Mater. Sci.*, **12** (1977), 2511.
78. Gettings, M. and Kinloch, A. J., *Surface Interf. Analysis*, **1** (1980), 189.
79. Lewis, B. F., Bowser, W. M., Horn, J. L., Luu, T. and Weinberg, W. H., *J. Vac. Sci. Technol.*, **11** (1974), 262.
80. Kerr, C., MacDonald, N. C. and Orman, S. *J. Appl. Chem.*, **17** (1967), 62.
81. Orman, S. and Kerr, C., in *Aspects of Adhesion—6*, ed. D. J. Alner, University of London Press, London (1971), p. 64.
82. Sharpe, L. H., *Appl. Polym. Sympos.*, **3** (1969), 353.
83. Cotter, J. L., in *Developments in Adhesives—1*, ed. W. C. Wake, Applied Science Publishers, London (1977), p. 1.
84. Zhurkov, S. N. and Korsukov, J. E., *J. Polym. Sci. Polym. Phys.*, **12** (1974), 385.
85. Levi, D. W., Wegman, R. F., Ross, M. C. and Garnis, E. A., *SAMPE J.*, **7**(3) (1971), 1.
86. Gledhill, R. A., Kinloch, A. J. and Shaw, S. J., *J. Adhesion*, **11** (1980), 3.
87. Kaelbe, D. H., *J. Appl. Polym. Sci.*, **18** (1974), 1869.
88. Owens, D. K., *J. Appl. Polym. Sci.*, **14** (1970), 1725.
89. Gettings, M., Baker, F. S. and Kinloch, A. J., *J. Appl. Polym. Sci.*, **21** (1977), 2375.
90. Minford, J. D., *Adhesives Age*, **20**(9) (1977), 41.
91. Bishop, H., Smart, N. R. and Kinloch, A. J., *J. Adhesion*, **14** (1982), 105.
92. Riel, F. J., *SAMPE J.*, **7**(4) (1971), 16.
93. McMillan, J. C., in *Developments in Adhesives—2*, ed. A. J. Kinloch, Applied Science Publishers, London (1981), p. 243.
94. Shannon, R. W. and Thrall, E. W., *J. Appl. Polym. Sci. Appl. Polym. Sympos.*, **32** (1977), 131.

95. Noland, J. S., in *Adhesion, Science and Technology*, ed. L. H. Lee, Plenum Press, New York (1975), p. 413.
96. Venables, J. D., McNamara, D. K., Sun, T. S., Ditchek, B. and Chen, J. M., in *Structural Adhesives and Bonding*, Technology Conf. Ass., El Segundo (1979), p. 12.
97. Sun, T. S., McNamara, D. K., Ahearn, J. S., Chen, J. M., Ditchek, B. and Venables, J. D., *Appl. Surf. Sci.*, **5** (1980), 406.
98. Kinloch, A. J. and Smart, N. R., *J. Adhesion*, **12** (1981), 23.
99. Sun, T. S., Chen, J. M., Venables, J. D. and Hopping, R., *Appl. Surf. Sci.*, **1** (1978), 202.
100. Kinloch, A. J., in *Adhesion—6*, ed. K. W. Allen, Applied Science Publishers, London (1982), p. 95.
101. Comyn, J., Horley, C. C., Oxley, D. P., Pritchard, R. G. and Tegg, J. L., *J. Adhesion*, **12** (1981), 171.
102. Allen, K. W., Alsalim, H. S. and Wake, W. C., *Faraday Spec. Disc.*, **2** (1972), 38.

Part I
FUNDAMENTAL ASPECTS

2

Surface and Interfacial Analysis

G. D. Davis and J. D. Venables

Martin Marietta Laboratories, Baltimore, Maryland, USA

NOTATION

AES	Auger electron spectroscopy
AMP	amino methyl phosphonic acid
EDS	energy dispersive spectroscopy
ESCA	electron spectroscopy for chemical analysis, synonym for XPS
FPL	Forest Products Laboratory pretreatment
FT-IR	Fourier transform infrared spectroscopy
HMP	hydroxy methyl phosphonic acid
IETS	inelastic tunnelling spectroscopy
PAA	phosphonic acid anodising
NTMP	nitrilotris methylene phosphonic acid
SEM	scanning electron microscopy
SIMS	secondary ion mass spectroscopy
WDS	wavelength dispersive spectroscopy
XES	X-ray emission spectroscopy
XPS	X-ray photoelectron spectroscopy
XSEM	high-resolution scanning electron microscopy

1. INTRODUCTION

Analysis of the properties of surfaces is becoming increasingly important in general studies of adhesive bonding, as well as in failure analysis. The chemical properties of a surface are frequently different from those of the bulk, and the morphological properties are always different. These differences result, in general, from the asymmetry of the sample caused by the surface, and more specifically, from oxide formation, segregation of alloying components, incorporation of electrolytic ions, adsorption of molecules from gases and liquids, and chemical reactions between the sample and its environment.

Many of these properties change dramatically from the first monolayer of atoms (3–4 Å) on the surface to the second, but others may change only over thousands of Ångstroms. Accordingly, the definition of a surface depends on what properties are being measured. In this chapter we consider the surface to include only the first few atomic layers, unless otherwise stated.

The chemical properties of an adherend surface govern the interaction of the adherend with a primer or adhesive as well as the long-term resistance of the bond to environmental degradation. The morphology, on the other hand, governs the degree of physical interlocking between the adherend and the primer or adhesive so that a microscopically rough surface can provide a stronger (and more durable) bond than a smooth one.

A large number of techniques exist for determining the properties of surfaces. Instrumentation for most of the major techniques is commercially available, and obtaining the appropriate measurements is becoming routine.

Many of these analytical tools have been applied to adhesive bonding problems. Each technique provides different, but complementary, information. Consequently the best approach toward either basic research or failure analysis is usually one that involves several techniques.

In this chapter we first discuss many of the surface analytical techniques applicable to the study of adhesive bonding. We concentrate on three techniques that we have found to be most generally useful in measuring the chemical and morphological properties of surfaces: X-ray photoelectron spectroscopy (XPS), Auger electron spectroscopy (AES), and high-resolution scanning electron microscopy (XSEM). Other techniques that have been used in investigations of adhesive bonding are covered in less detail. For more information on any of the analytical techniques, or on surface science in general, the reader is referred to the bibliography.

Drawing on the work done at Martin Marietta Laboratories, we then discuss applications of many of these techniques. The discussion is organised in terms of surfaces investigated rather than the analytical techniques themselves, in order to illustrate the complementary results obtained from using several techniques and to demonstrate that a complete understanding of a problem is often difficult, if not impossible, using a single method.

2. TECHNIQUES FOR SURFACE ANALYSIS

The ideal technique for surface chemical analysis should provide a qualitative and quantitative elemental analysis of the first monolayer of the surface and identify the type of bonding present at the surface. In addition, the measurement process should not alter the surface, it should probe only a small area of the surface to resolve regions of inhomogeneity; it should have near-uniform high sensitivity for all elements; and it should provide, or at least permit, a profile into the near-surface ($<1\ \mu$m) region of the sample. No technique, of course, will satisfy all of these requirements, but two or three techniques can frequently be combined to obtain a satisfactory analysis.

2.1 X-ray Photoelectron Spectroscopy

X-ray photoelectron spectroscopy (XPS), also known as electron spectroscopy for chemical analysis (ESCA), can provide very good qualitative and quantitative elemental analysis of a surface and reasonably good determination of the chemical bonding. XPS involves the bombardment of the sample surface with mono-energetic X-rays in ultrahigh vacuum ($<10^{-7}$ Torr, $<10^{-5}$ Pa). As the photons (of energy $h\nu$) travel through the material, some are absorbed and their energy is transferred to electrons which can be ejected from the sample. Conservation of energy requires that:

$$E_k = h\nu - E_B \qquad (1)$$

where E_k is the kinetic energy of the photoelectron and E_B is the energy, relative to the vacuum level, that binds the electron to the atom. The spectrum, the number of electrons per unit energy $N(E)$ versus binding energy, is obtained by pulse-counting techniques. It consists of a series of peaks superimposed on a sloping background (Fig. 1).

The surface sensitivity of XPS is due to the strong interactions that the photoelectrons [with typical kinetic energies of 200–1200 eV after excitation with Mg (1253·6 eV) or Al (1486·6 eV) Kα radiation] have with the other electrons of the sample. As a result, only photoelectrons originating very near the surface (10–30 Å) have a high probability of

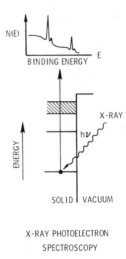

X-RAY PHOTOELECTRON
SPECTROSCOPY

Fig. 1. Schematic representation of X-ray photoemission process and typical spectrum. An X-ray photon is absorbed by an atom which emits a photoelectron.

being emitted into the vacuum without loss of energy. (Those electrons that have lost energy form the sloping background of the spectrum and generally do not provide useful information.)

The peaks in the spectrum represent the core levels of atoms present on the surface. Since each element has core levels at a unique set of energies, qualitative analysis is achieved by comparing the peak energies with values obtained from standards or from tables.[1,2] In principle all elements are detectable, but in practice hydrogen is not seen since its electron becomes involved in the bonding and has no characteristic energy.

The observed binding energies of electrons from a given core level can change for two reasons. The first is that charging results in a uniform shift to higher binding energy (≤ 15 eV) for all the peaks in a spectrum of a surface that is not in good electrical contact with ground. As electrons are emitted, the surface becomes positively charged; the resulting electric field slows the photoelectrons (thus shifting the peaks to higher binding energies). This effect can be compensated by assuming that the binding energy of the 1s level of the adventitious carbon is 284·6 eV and shifting the energies of all peaks accordingly. Alternatively, the surface can be flooded with low-energy electrons so that the charge on the surface is neutralised; the peaks will then appear at the correct binding energies.

The second case in which changes in binding energies occur is when the peaks of only one element are shifted ($\leqslant 5$ eV), reflecting a change in the chemical state of the element. As electronic charge is transferred from (to) an atom, the screening it provides of the positively charged nucleus decreases (increases) and the binding energies of the core levels increase (decrease). Hence, the chemical shift of a photoelectron line, relative to that of the elemental form or other known standard, can be compared to the experimentally or theoretically determined chemical shifts of other compounds[1,2] to help identify compounds on the surface.

For example, the chemical shifts of the Si $2p$ and O $1s$ lines have made it possible to identify silicon-containing contamination on adherend surfaces as either silicones or silicates. To eliminate the effect of charging and to avoid reference to the C $1s$ signal, which is the sum of the signal from the epoxy and those from adventitious hydrocarbons, the binding energy difference between the O $1s$ and the Si $2p$ peaks was obtained. By comparing the results from the unknown with those of standards (431·0 eV for silicates and 432·7 eV for silicones), the source of the contamination could be determined.

The chemical shifts of photoelectron lines are not unique, however, and it may not be possible to determine with certainty the compounds present from these data alone. In such cases Auger peaks (described in section 2.2) that appear in the XPS spectra of many elements may be helpful. Additional compounds can be identified[1,3–6] by comparing modified Auger parameters, α', defined as:[3,4]

$$\alpha' = E_k(A) - E_B(P) \tag{2}$$

where $E_k(A)$ is the kinetic energy of the Auger electrons and $E_B(P)$ is the binding energy of the photoelectrons.

Quantitative analysis is obtained from the areas under the major peak of each element or, alternatively, from the height of these peaks. The number of photoelectrons $N(E)$ emitted from a given core level and measured by an analyser with a resolution dE is given by:

$$N(E)\,dE = \sigma T(E) \int_0^\infty I_{h\nu}(z) n(z) e^{-z/\lambda}\,dz\,dE \tag{3}$$

where σ is the photoionisation cross-section for a given core level of a

given element, $T(E)$ is the transmission of the electron energy analyser, $I_{h\nu}(z)$ is the photon flux at a depth z in the sample, $n(z)$ is the
atomic density at depth z of the element of interest, and λ is the mean
free path of the photoelectrons in the solid. This expression can be
simplified by noting that $I_{h\nu}(z)$ is essentially constant over the depth
from which electrons are measured and can be factored from the
integral. If the sample composition is assumed uniform over the
sampling depth (an assumption that may not be valid), the integration
can be performed to obtain:

$$N(E)\,dE = \sigma T(E)I_{h\nu}\lambda n\,dE \qquad (4)$$

The quantities σ, $T(E)$ and λ are constant for a given photoelectron
line and a given analyser (possible variations of λ with different
materials are usually ignored) and can be combined into a sensitivity
factor, S, that can be calculated or measured from standards. Several
compilations of sensitivity factors have been published,[1,7,8] but care
must be taken to ensure that the $T(E)$ appropriate to the analyser used
was included in the tabulated values. For best results, accurate sensitivity factors for elements of great interest should be measured from
standards in individual laboratories. Finally, effects of the variations in
photon flux from sample to sample are eliminated by summing over all
elements and normalising to 100%, so that the concentration C_j of
element j is:

$$C_j = \frac{\dfrac{N(E_j)}{S_j}}{\displaystyle\sum_j \dfrac{N(E_j)}{S_j}} \times 100\% \qquad (5)$$

The sensitivity factors for most common elements vary by approximately one order of magnitude from element to element, with an
average detectability limit of $\sim 0.2\%$ of the sampling region or ~ 0.5–
1% of a monolayer, depending on the data acquisition time.

It is often necessary to determine the changes of surface composition
during a chemical reaction. For surfaces and reactions with a limited
number of constituents and reaction products, a convenient way to
trace these changes is the construction of a surface behaviour diagram,[9]
an example of the use of which will be described in detail in section
3.2 of this chapter. These diagrams are similar in appearance to phase
diagrams for equilibrium of bulk compositions, but they present sur-

face compositions with the equilibrium condition relaxed. A limited number of compounds (usually three or four) are judiciously chosen so that the surface composition at any point in the reaction can be expressed as a linear combination of the compositions of these compounds. (For example, compositions corresponding to AlOOH and $Al(OH)_3$ can be expressed as linear combinations of the compositions of H_2O and Al_2O_3 although the compounds themselves are not strictly mixtures of H_2O and Al_2O_3.) The compositions of the basic compounds are then assigned to the vertices of an equilateral triangle or a tetrahedron. By tracing the surface composition at different stages of the reaction on a surface behaviour diagram, the reaction path can be compared to those predicted by different proposed reactions, and a model can be developed to explain the data.

A similar construction can also be used to trace the composition of a sample as a function of depth.[10] It readily allows small changes in the composition of the oxide or other overlayer to be detected and portrays them in such a way that the interaction of the overlayer with the substrate can be examined in detail.

2.2. Auger Electron Spectroscopy

Auger electron spectroscopy (AES) is frequently used as a complement to XPS for surface analysis. Quantitative and chemical-state analyses are not as good as those of XPS, but AES in combination with inert-ion sputtering can provide depth profiles and is also capable of good lateral resolution. AES usually involves the bombardment of the sample surface, in ultrahigh vacuum, with 2–10 keV electrons which ionise some of the atoms, creating a core hole (Fig. 2). (Core holes are also created by photoionisation, so that Auger transitions appear in XPS spectra also, as mentioned in section 2.1.) The energy of this ion is reduced when an electron in a higher level drops down to fill the core hole. To conserve energy, a second electron is emitted from the atom, leaving it doubly ionised. The kinetic energy $E_k(WXY)$ of an Auger electron from transition WXY is approximated by:

$$E_k(WXY) = E_B(W) - E_B(X) - E_B(Y) \qquad (6)$$

where $E_B(W)$ is the binding energy of an electron in core level W (which had the initial core hole) and $E_B(X)$ and $E_B(Y)$ are the corresponding quantities for the 'down' and 'up' electrons in levels X and Y, respectively (Fig. 2). [More accurately, $E_B(X)$ and $E_B(Y)$ are the binding energies of electrons in levels X and Y in the presence of a

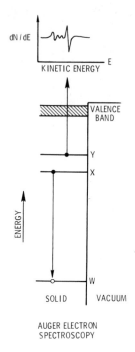

Fig. 2. Schematic representation of a WXY Auger transition and typical spectrum. A core hole in the W level is filled by an electron from the X level which causes an electron from the Y level to be emitted.

hole in level W.] The energy of an Auger electron is determined only by the binding energies of the electrons involved in the transition (unlike that of a photoelectron), and it is therefore independent of the ionising radiation.

Because an Auger transition requires two electrons above the lowest level, H, He, and gaseous Li exhibit no Auger transitions. Each of the remaining elements possesses a unique spectrum with one or more Auger peaks, and compilations of spectra[11] or calculated energies[12] are available. Like XPS, AES derives its surface sensitivity from the strong interaction of low kinetic energy electrons with the material.

Traditionally, an Auger spectrum is taken as dN/dE vs E to accentuate the small, sharp Auger peaks which are superimposed on a large, slowly varying background of inelastically scattered electrons. This spectrum is obtained using standard phase detection techniques.

(There is a recent trend to pulse-count the Auger electrons and to obtain an $N(E)$ vs E spectrum, but data are frequently still reported in the dN/dE mode.)

The number of Auger electrons $N(E)$ emitted from a given transition and measured by the analyser can be expressed as:

$$N(E)\,dE = \sigma(E_p)\alpha[1+\nu]T(E)B(E_p)\int_0^\infty I_p(z)n(z)e^{-z/\lambda}\,dz\,dE \quad (7)$$

where $\sigma(E_p)$ is the ionisation cross-section for the given initial core hole and for the given primary energy E_p; α is the probability that the core hole will decay with the given Auger transition; ν is the probability that additional holes in the initial core level are created as a result of other Auger transitions; $B(E_p)$ is a correction for backscattered primary electrons; $I_p(z)$ is the primary electron flux at a depth z in the sample; and other terms are as defined in eqn (3). If we make the same simplifying assumptions as previously, i.e. that $I_p(z)$ is constant over the depth from which electrons are measured (which is valid as long as $E_p \gg E$) and that the surface composition is uniform over the same depth, we obtain:

$$N(E)\,dE = \sigma(E_p)\alpha[1+\nu]T(E)B(E_p)I_p(z)\lambda n\,dE \quad (8)$$

The terms multiplying n can be combined into a sensitivity factor and the surface composition determined by normalising the total concentration to 100% as discussed in connection with XPS. However, the standard practice of measuring peak-to-peak heights of Auger lines in the dN/dE mode, instead of peak areas in the $N(E)$ mode (i.e. the second integral of the dN/dE spectrum), makes quantitative analysis more difficult with AES than with XPS.

Peak-to-peak heights are proportional to the number of Auger electrons emitted only if line shape of the transition is constant—a condition that does not always hold. Several elements, including some important constituents of adherends, such as Al, Ti, and Mg, exhibit Auger transitions whose energies and line shapes differ for the metallic and oxidised states.[13] Accordingly, the peak-to-peak height sensitivity factor for the metal can differ from that for the oxide. Furthermore, the sensitivity factor for a mixture can be a non-linear function of those of the two components and quantitative analysis may be difficult to obtain from Auger depth profiles of such systems.

In some cases it is possible to use the changes in the energy or line shape of a transition[13,14] or changes in the electron inelastic scattering to determine chemical information.[15] This is especially true if one or two of the electrons involved in the transition originate in the valence band. The energy distribution of these Auger electrons reflect the density of electronic states in the valence band[13,14] and can be used as a fingerprint to identify different compounds. For example, the line shape of the Ti LMV transition (LMN when the N electrons are in the valence band) is different for elemental Ti, TiO, Ti_2O_3 and $TiO_2^{16,17}$ and can therefore be used as a measure of the stoichiometry of Ti oxides.

Although XPS has been shown to have many advantages over AES in the comparisons presented up to this point, AES has the capability of providing depth profiles and elemental maps with high lateral resolution that are difficult, if not impossible, to obtain with XPS. Of the two advantages, the former is more important for adhesive bonding applications and examples will be presented in section 3.

By combining AES with inert-ion sputtering, one can measure the composition of the surface as material is milled away to expose underlying atomic layers. This procedure allows elemental concentrations to be determined (within the limitations discussed above) for the first few hundred nanometres of a sample surface and is especially useful for obtaining the thicknesses of thin overlayers. The sputtering rate for removal of material must be calibrated for each spectrometer for one standard material. Compilations of the relative sputter yields for different materials are available[18] to convert, in principle, the sputtering rate of the standard to those of other samples. In practice, however, it is difficult to obtain good values of sputtering rates for many materials, but if we seek only differences in the thickness of overlayers, absolute sputtering rates are not needed; variations in the time to sputter through the overlayer are sufficient.

It is important to note that the sputter time is proportional to the amount of material removed, not to the actual thickness as measured in an electron microscope. Thus, for example, if the sputtering rate is calibrated for a dense oxide, the apparent thickness (obtained from the Auger depth profiles) of a porous oxide will be less than the actual thickness. The difference between the two measurements will be an indication of the degree of porosity of the oxide.[19]

The other important capability of AES is the possibility of high lateral resolution. By electrostatically focusing and rastering the electron beam, elemental maps with a resolution of up to 500 Å can be obtained from sample surfaces. Although scanning Auger microscopy

(SAM) has found much use in the semiconductor industry due to its ability to map the different metallisations and other deposited or grown coatings, it has not found a similar degree of use to date in studies of adhesive bonding. This lack of application is partially due to the inherent lateral homogeneity of the adherend surfaces (within the resolution of SAM) and partially to difficulties in examining polymers (and insulating surfaces in general) with an electron beam.

For most materials and primary electron beam energies, the secondary electron emission yield is less than unity, i.e. fewer electrons are ejected from the surface than are incident upon it. This creates no problem if a conductive sample is examined; electrons flow to ground and maintain the charge neutrality of the sample. However, if the sample is insulating or if it is not grounded, the sample surface will charge negatively. In electron-microscopic studies, such a sample is coated with a thin metallic layer to conduct the charge to ground. Properly deposited, this layer does not alter the morphology within the resolution of the microscope. This solution is not possible for AES, however, as the overlayer would mask the signal of the substrate. In cases of mild charging, the peaks of the Auger spectrum are all shifted uniformly to higher kinetic energy, but in severe cases, a spectrum cannot be obtained. If charging occurs, it can sometimes be reduced or eliminated by reducing the electron beam current, by raising or lowering the beam voltage, by reducing the angle of incidence of the electron beam (making it more glancing), or by light inert-ion-sputtering the surface to provide a source of positive charge. The last procedure, of course, cannot be used if a very thin layer of surface contamination is of interest.

In addition to charging, polymeric surfaces (and some others) are often subject to electron-beam-induced damage which can preclude the use of AES. This damage can be either dose-related, in which case it is due to electron excitation, or flux-related, in which case it is due to heating. Both mechanisms result in broken chemical bonds and decomposition of the polymer. Occasionally the damage can be lessened by reducing the electron beam current and rastering the beam, but in most cases XPS or other techniques must be used.

2.3. Other Techniques

Although XPS and AES are the most common and perhaps the most versatile tools of surface analysis, other techniques are available that can provide additional information. Some of these complementary techniques are discussed briefly below:

2.3.1. Secondary Ion Mass Spectroscopy (SIMS)

This technique detects ions ejected from a surface by momentum transfer from an incident ion beam. It is capable of detecting all elements (including hydrogen), allows isotope discrimination, and has very good sensitivity for some elements (100 ppm of a uniformly distributed material). The surface sensitivity results because the ejected ions originate in the top few layers of the solid. SIMS allows mass analysis of both positive and negative ions.

The number $N^+(A)$ of positive ions of isotope A detected by the analyser can be written as:

$$N^+(A) = \alpha^+(A)n(A)SIT \tag{9}$$

where $\alpha^+(A)$ is the ratio of positive secondary ions of A to the total number of sputtered A atoms, $n(A)$ is the number of A atoms, S is the total sputtering yield of A atoms, I is the incident ion current, and T is the transmission of the analyser. An analogous expression exists for negative ions. Quantitative analysis is very difficult with SIMS because $\alpha^+(A)$ and S are very material-dependent, i.e. $\alpha^+(A)$ is sensitive to the electronic configuration and S is sensitive to the relative cohesive energies. In addition, they are also functions of the incident ion so that the ion source can be chosen to emphasise positive or negative emitted ions. Thus published sensitivity factors[20] are only typical values, and even calibration with standards may not be adequate if the material matrix changes.

Chemical information is potentially available because ion clusters are also detected. Interpretation, however, can be difficult because the clusters may have been ejected from the sample or formed in the vacuum above it. Identification may also be difficult with an unknown sample since only the mass of the ion can be determined.

Unlike XPS and AES, SIMS is by its very nature destructive. The detection limit cannot be improved by longer data acquisition times; instead, a minimum number of atoms are required for analysis. Consequently, although imaging is possible with SIMS, it is at the expense of depth resolution. Low concentrations in a thin surface layer are best detected using a slow sputtering rate and a spatially large incident beam—the so-called 'static' mode of detection. Imaging, on the other hand, requires a fast sputtering rate, i.e. the 'dynamic' mode of detection.

2.3.2. *Inelastic Electron Tunnelling Spectroscopy (IETS)*

This is used to measure the vibrational modes of molecules adsorbed on to metallic, usually aluminium, oxides and provides information concerning the chemical bonding of the adsorbate and the substrate. A sample is prepared by adsorbing, either from vapour or liquid, a monolayer of the molecule of interest on to an aluminium substrate covered with approximately 20 Å of oxide. A lead film is then evaporated on to the oxide plus adsorbate. The oxide film acts as a barrier to current flow; however, if a potential V is applied across the metallic layers, some electrons are able to tunnel through the oxide (Fig. 3). A small fraction of the tunnelling electrons excites vibrational modes of the adsorbate and loses energy. At the thresholds of these excitations, the current–voltage curve changes slope. As with AES, phase detection is used to accentuate the thresholds; in this case the spectrum is obtained as dI^2/dV^2 vs V. To improve the signal-to-noise ratio (S/N), the junction is usually cooled to liquid helium temperatures (4·2 K) or below. This allows investigation of sub-monolayer coverages to be investigated—a significant improvement over other vibrational spectroscopies such as infrared and Raman.

The energies of the different peaks in the spectrum can be identified with the vibrational modes of the different chemical bonds. These assignments permit the bonding between the adsorbate and the substrate to be determined, while the relative intensities of the peaks may permit the orientation of the molecule to be established.

2.3.3. *Fourier Transform Infrared Spectroscopy (FT-IR)*

Fourier transform infrared spectroscopy also provides a vibrational spectrum, using optical interferometric techniques. A Michelson interferometer provides a Fourier transform of the infrared spectrum transmitted, reflected or emitted by the sample. A minicomputer performs the inverse transform back to the frequency spectrum.

If we compare the energies of the vibrational modes in a spectrum with those derived by calculations or from standards, the chemical bonds of a sample may be identified. Consequently, more information concerning the chemical bonding of organic or inorganic materials is available from FT-IR than can be obtained from XPS.

Because photons rather than free electrons are used to probe the sample, FT-IR does not require a high vacuum and can be used to examine many materials, including solvents, that are not compatible with high vacuum. On the other hand, since photons have long mean free paths in solids, FT-IR is not very surface-sensitive. Data taken in

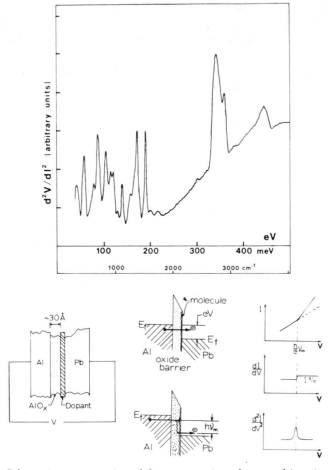

Fig. 3. Schematic representation of the cross-section of a tunnel junction, elastic and inelastic electron tunnelling processes, and current–voltage features near $eV = h\nu_m$ where ν_m is the frequency of a mode. Also shown is an IETS spectrum of benzoic acid adsorbed on aluminium oxide.[21]

the transmission mode are averaged over the entire sample. Measurement in the reflection mode improves the surface sensitivity to 1–3 μm; this can be improved, in some cases, to less than 100 Å by deposition of the sample material on to a vaporised metal mirror.

The extreme surface sensitivity of the electron and ion spectroscopies is not always necessary, however, and FT-IR has proved useful in the characterisation of primers and adhesives and in failure analysis. When high surface sensitivity is needed, surface contaminants can sometimes be extracted in an appropriate solvent and concentrated before analysis. Spectra of the solution, both with and without the contaminant, are then compared to identify the contaminant. This technique can detect and identify even a few nanograms of material, which correspond to approximately one monolayer on a 1 cm^2 sample. In general, however, FT-IR is best suited for determining the types of chemical bonds present in thin (1–3 μm) organic and inorganic layers on surfaces.

2.3.4. Scanning Electron Microscopy (SEM)

For a better understanding of many adhesive bonding problems, the chemical information obtained with the techniques presented above must be complemented with the morphological information obtained from scanning electron microscopy. Due to the scale of the microstructures ($\sim$50 Å) present on adherend surfaces (to be discussed in section 3), ultrahigh resolution (XSEM) is required. Such resolution is only available on scanning transmission electron microscopes (STEM) or on scanning electron microscopes utilising the same principles.

Two factors are important in achieving this ultrahigh resolution. The first arises because the sample is placed within the objective lens. The resulting short focal length minimises the spherical aberration and permits a smaller beam size.[22] The resolution is further increased for a given beam size by using a magnetic field as a momentum filter to allow detection of only the secondary electrons. These electrons, with kinetic energies less than 50 eV, originate in the first few hundred Ångstroms of the sample, before the excitation volume has expanded to the classic teardrop shape.[23] Consequently, the area from which secondary electrons are detected is only slightly larger than the probe diameter, in contrast to the area from which backscattered electrons are detected. (These electrons, which occur at higher kinetic energy, originate, on average, much deeper in the sample, after the expansion of the excitation volume. Accordingly, the area from which they are detected is much larger, approximately five times the beam diameter.)[23]

At this resolution, the sputtered or evaporated gold coatings (normally used in microscopy for reducing the charging of insulating materials that was discussed in connection with Auger electron spectroscopy (section 2.2)) are non-uniform and exhibit structure that disguises the true surface oxide morphology. Instead, very thin

(~50 Å) platinum or gold–palladium films are sputtered on to the oxide. These coatings are sufficient to prevent charging of the surface, but introduce no significant structure at magnifications of 50 000.

Conventional micrographs, by their very nature, are two-dimensional; using them to interpret three-dimensional surface features can be difficult and even misleading. However, by taking micrographs of the same area both before and after tilting the sample a few degrees, a stereo pair is obtained.[24] An observer examining the two micrographs through a stereo viewer will use the parallax to form a three-dimensional view of the sample. Such micrographs have proven very useful in the studies of adherend surfaces that are reported in section 3.

A companion technique to the imaging provided by most SEM is X-ray emission spectroscopy (XES). It provides elemental analysis by analysing either the wavelength or energy of X-rays emitted from the samples as a result of a decay mechanism in competition with an Auger transition. (The initial core hole is filled by an electron in an upper level, as in the Auger process, but the excess energy is used to create a photon instead of ejecting a second electron.) Energy dispersive spectroscopy (EDS) involves the detection of X-rays with a Si(Li) solid state detector. It is the more common and faster procedure for detection of X-rays but has much poorer resolution and hence is more difficult to quantify. Wavelength dispersive spectroscopy (WDS), on the other hand, involves the use of crystals to diffract the X-rays. It is limited by the need to move the crystal mechanically and to use several different crystal spectrometers in order to scan a wide range of X-ray wavelengths. Because of the poor detection efficiency for low-energy X-rays, and the low probability that a shallow core hole (binding energy <2 keV) will decay by X-ray emission, light elements thave small XES sensitivity factors and the lightest ones cannot be detected at all. (EDS is usually limited by the window of the detector and cannot detect elements lighter than Na, whilst WDS and EDS using windowless detectors can detect elements as light as C or B.) Surface sensitivity is governed by the mean free path of the high-energy electrons used to excite the core hole and hence is poor (~ 1 μm).

2.3.5. Ellipsometry
Ellipsometry characterises the light reflected from a sample; in principle, it can be used to measure the thickness and other properties of an oxide or other overlayer.[25] These measurements require an automatic

spectroscopic ellipsometer which is microprocessor-controlled. If, on the other hand, one is interested only in changes of the thin film, these can be detected using a simpler null ellipsometer.

An ellipsometer has three major optical components: a polariser, a compensator and an analyser. The compensator converts the incoming plane-polarised light into elliptically polarised light characterised by two perpendicular energy vectors that differ in phase. In a null ellipsometer, the polariser is used to adjust this phase difference so that it exactly compensates for the phase shift that occurs upon reflection from the sample. The analyser is then used to extinguish the plane-polarised beam reflected from the sample surface. The point of extinction represents a null point which is specific for the thickness and optical constants of a film on the sample surface. Thus, any subsequent change in the film results in an observed signal. For example, if the null point is set for an oxide surface, any hydration or oxide growth will be detected. Consequently, an incubation time prior to hydration can be determined. Examples of such an application are discussed in section 3 of this chapter and are referred to on many occasions in subsequent chapters.

3. APPLICATIONS

In this section we present examples of the application of some of the surface analytical techniques described earlier. Because complementary information derived from several techniques is often required for a complete understanding of a problem, the investigations discussed here were chosen to illustrate the use of several different analytic tools and their interplay.

3.1. Forest Products Laboratory Pretreatment for Aluminium Adherend Surfaces

The Forest Products Laboratory (FPL) process has been used for many years to prepare aluminium surfaces for fabrication into adhesively bonded aircraft structures.[26] It consists of a number of cleaning, etching, and rinsing steps. The pre-etch preparation includes vapour degreasing with inhibited 1,1,1-trichloroethane and alkaline cleaning with Turco 4215-S cleaner, followed immediately by rinsing in water. The etch itself, at 150°F (65°C) for 30 min, is in an aqueous solution of H_2SO_4 and $Na_2Cr_2O_7$. The etch is followed by another series of rinses.

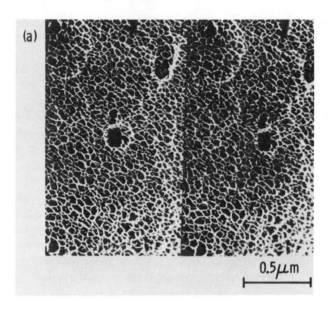

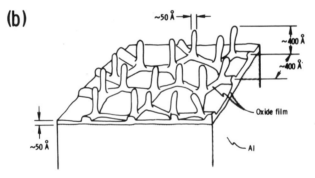

Fig. 4. (a) *Ultra-high resolution stereo XSEM micrograph and* (b) *isometric drawing of the oxide morphology on an FPL-treated aluminium surface* (50 000×).[27]

An XSEM stereo pair of the surface morphology of an FPL etched coupon is shown in Fig. 4, accompanied by an isometric drawing of the oxide structure. An identical morphology is seen in stripped oxide films with TEM stereo pairs.[27] The oxide is characterised by a cell structure and a high concentration ($\sim 10^{10}/cm^2$) of 50Å-thick, 400Å-high oxide whiskers that protrude from the surface. These protrusions, which are not evident from a single, two-dimensional micrograph, may mechanically interlock with the adhesive to provide good bond strength.

To characterise the FPL surface further and to explore possibilities for improving the resulting adhesive bonds, a series of investigations on its growth, its hydration, and the inhibition of this hydration have been undertaken.[28-31]

Auger spectroscopy and XPS of the FPL oxide on 2024 aluminium revealed the presence of only Al, O, Cu, and C (Fig. 5). Based on the binding energies of the XPS lines and the corresponding quantitative analysis, the surface was identified as Al_2O_3 with adsorbed H_2O and ever-present adventitious hydrocarbon contamination. The Cu was present as a constituent of the alloy and as a Cu-rich layer found between the oxide and metal (Fig. 6(b)).

The depth profile, together with the calibrated sputter rate, showed that the FPL oxide had an equivalent dense oxide thickness of 100 Å. The apparent decrease in the Al concentration at ~7 min was not real, but was an artifact of a chemical shift of the Al KLL line. As the Al oxide peak decreased and the Al metal peak increased, they partially interfered destructively causing a decrease in the peak-to-peak height.

A comparison of this depth profile with one taken of a similar

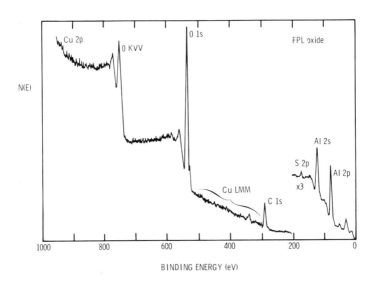

Fig. 5. *XPS spectrum of FPL surface.*

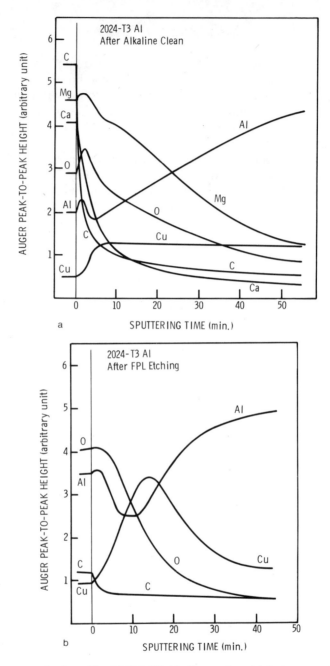

Fig. 6. *Auger depth profile of 2024-T3 Al alloy surface (a) before, and (b) after FPL etch. Multiplication factors: Al × 1; Mg, Cu, Ca, C × 2; O × $\frac{1}{4}$.*[28]

coupon before etching (Fig. 6(a)) showed that the sulphuric acid–sodium dichromate etch removed the original oxide and the surface contamination (MgO from the heat treatment and Ca from the alkaline etch). It further showed that the Cu accumulation at the oxide/metal interface occurred as a result of the etch. Similar measurements on 7075-T6 Al (which has less Cu) also showed a Cu accumulation, but to a correspondingly lesser extent,[28] suggesting that the Cu originates in the bulk and not in the etching solution.

These results and others[29] allowed a growth mechanism to be proposed. During the etching process, both formation of new oxide and dissolution of existing oxide take place. At steady state, the rates of these two processes become equal, but a thin oxide layer is maintained at the surface. As etching continues, the boundary of the oxide retreats from the solution interface, and the material is dissolved at such a rate that the proportion of the alloying components dissolved equals the composition of the bulk. Since oxidation involves diffusion of ions, it appears that Al and the alloying ions diffuse through the oxide layer into the etching solution, while the boundary of the oxide is retreating. The build-up of Cu at the oxide–metal interface appears to imply a low diffusion rate of Cu through the oxide compared with that of Al ions.

Although it is the rough morphology of FPL oxide that determines the high initial strength of the joint, it is the stability of the oxide in humid environments that determines its long-term durability. This became evident during the examination of failed surfaces from aluminium Boeing wedge-test samples (see Chapter 1, Fig. 5) after exposure to accelerated test conditions (100% relative humidity at 65°C, and a water-wicking adhesive with no primer). The crack, which initially propagated through the adhesive before exposure to humidity, immediately travelled along the (macroscopic) adhesive/oxide interface upon exposure. High-resolution stereo micrographs showed a new 'cornflake' morphology (Fig. 7) on both sides of the failure while XPS and electron diffraction identified this hydration product as crystallites of boehmite AlOOH. Auger depth profiles, however, revealed a significant difference between the two sides. The hydroxide on the adhesive side of the failure was approximately three times the original FPL thickness while that on the metal side was much thinner. These results strongly suggest that the joint failure occurred at or near the boehmite–metal interface. Apparently, the adhesion of the hydroxide to the aluminium was sufficiently weak, once the hydroxide formed, for

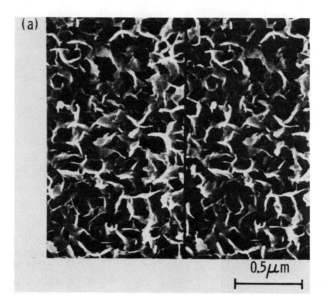

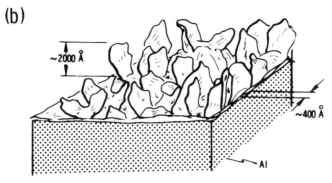

Fig. 7. (a) Ultra-high resolution stereo XSEM micrograph, and (b) isometric drawing of the aluminium oxyhydroxide 'cornflake' morphology produced during immersion in 80°C water (50 000×).[30]

it to separate from the substrate, giving rise to joint failure. The newly exposed Al surface then hydrated also.

If this proposed failure mechanism for adhesive joints in a humid environment is correct, then a correlation would be expected between hydration rates of oxide surfaces and wedge-test results. Such hydration rates can be found by ellipsometry. A typical measurement made on FPL treated 2024-T3 Al exposed to 80°C deionised water is shown in Fig. 8, along with XSEM stereo micrographs that depict the evolution of the morphology as a function of exposure time. The data and micrographs indicate that there was an incubation time of approximately 2 min during which the optical properties and appearance of

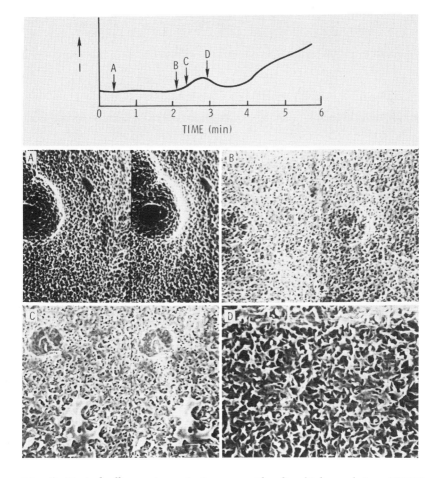

Fig. 8. *Typical ellipsometer output curve with ultra-high resolution XSEM micrographs of sample surfaces removed at various points along the curve: (a) the original FPL morphology; (b) the ellipsometer output just beginning to increase, and the surface morphology showing some filling in of pores; (c) the cornflake morphology beginning; and (d) the surface completely converted to hydroxide.*

the oxide changed very little. After this time, the oxide surface began to hydrate. First, the oxide porosity began to fill in (Fig. 8(b)); then the surface began to roughen considerably (Fig. 8(c)); and finally the characteristic 'cornflake' structure of boehmite developed (Fig. 8(d)) by the reaction:

$$Al_2O_3 + H_2O \rightarrow 2AlOOH \qquad (10)$$

After this point, gas evolution was detected, suggesting that the aluminium metal had come in contact with the water and was corroding by the reaction:

$$2Al + 4H_2O \rightarrow 2AlOOH + 3H_2 \uparrow \qquad (11)$$

Evidently, the conversion of the original oxide to a hydroxide may lead not only to the degradation of adhesive bonds but also to the general corrosion of aluminium.

Using the incubation time associated with the hydration process as a criterion for evaluating the stability of oxide surfaces, this value was measured for various types of surfaces and the results were compared with wedge-test data to determine whether there was a correlation.[30,31] In the first of these experiments, the incubation time for phosphoric acid anodised (PAA) surfaces was found to be considerably greater than the 2 min incubation time for FPL surfaces. A great deal of scatter was observed in the data with values ranging from 15 min to 16 h, but the most frequently observed times were in the 3–5 h range. This difference between incubation times for FPL vs PAA can be compared with wedge-test results[31,32] that showed the wedge-test performance of PAA-treated adherends to be far superior to that for FPL. These results suggested a possible correlation between the stability of surface oxides on Al and wedge-test performance.

To eliminate the possibility that morphological differences were responsible for the wedge-test results, further corroborating evidence was sought in studies originally intended to test the effect on wedge tests of exposing aluminium adherends to conditions used for curing high-temperature adhesives.[33] In these experiments, wedge-test results and incubation times for magnesium-containing aluminium adherends, such as 7075, improved as the time at high temperature (175°C) increased before bonding. Although XSEM examination of the treated surfaces indicated no detectable change in the FPL morphology, Auger analysis indicated a significant increase in the magnesium content of the oxide as the heat-treatment time increased. The improvement in wedge-test results was attributed to the increased Mg content of the oxide, although a dehydration effect could not be discounted. In any case, a correlation was found between the mechanical property data and surface hydration rates—the longer the incubation time, the better the wedge-test performance.

This correlation led to an investigation of methods to inhibit the hydration process. One method is to adsorb a thin layer of an inhibitor on to the FPL surface. One such family of inhibitors includes nitrilotris methylene phosphonic acid (NTMP), aminomethylphosphonic acid (AMP), and hydroxymethylphosphonic acid (HMP).[31] The structures of the deprotonated molecules are shown schematically in Fig. 9.

An XPS spectrum of an FPL surface after treatment with one of these inhibitors showed only the addition of N and P (and additional O and C) to the original surface. Stereo micrographs showed a morphology identical to the untreated surface.

To determine the amount of inhibitor adsorbed on an adherend surface, the peak heights of the $2p$ spectral lines of P and Al were measured and the P/Al ratio was calculated using the appropriate sensitivity factors.[9] The P/Al ratio was taken as the relative coverage of inhibitor on the aluminium oxide surface. Figure 10 summarises the results of this analysis as a function of inhibitor concentration. The

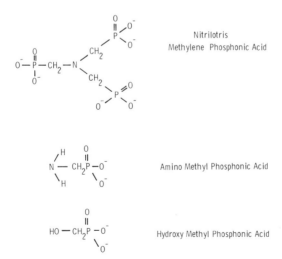

Fig. 9. Schematic of deprotonated inhibitor molecules.

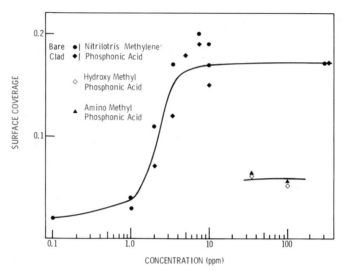

Fig. 10. Inhibitor surface coverage (P/Al ratio as determined by XPS) of FPL-etched 2024 Al as a function of inhibitor solution concentration.[31]

P/Al ratios for these molecules indicated approximately monomolecular coverage at saturation. The threefold increase in the P/Al ratio for NTMP compared to AMP and HMP suggested that only one leg of the NTMP molecule was attached to the surface.

As mentioned above, incubation time for conversion of oxide to hydroxide on FPL-etched coupons was approximately 2 min when no inhibitor was used. Treatment of the FPL surface with any of the inhibitors increased the incubation time considerably. For example, some coupons with saturation coverage of NTMP hydrated in 15 min while others did not hydrate at all after 23 h of immersion in water held at 80°C.

Because the incubation time measurements on coupons treated with inhibitors had considerable scatter, it is somewhat difficult to use the incubation time for quantitative comparison with wedge-test results. The experimental scatter may have been caused by inclusions in the 2024 Al samples, which acted as nucleation sites for hydration (Fig. 11). The sample shown in this Figure was removed from its 80°C water environment just as the ellipsometry trace showed signs of early hydration activity. When it was examined by SEM, circular patches of hydration product were observed. A chemical analysis performed by energy dispersive spectroscopy (EDS) revealed that at the centre of each hydration patch the Cu content was very high, approaching that of a $CuAl_2$ inclusion. It was speculated that the sites of the inclusions may be particularly difficult to render passive because of local galvanic currents and/or because the aluminium oxide film was discontinuous there. In any event, even though inclusions may have led to scatter in

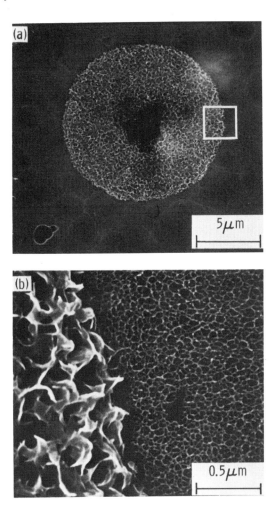

Fig. 11. SEM micrographs of a hydration 'island' surrounding a $CuAl_2$ inclusion at medium and high magnification.

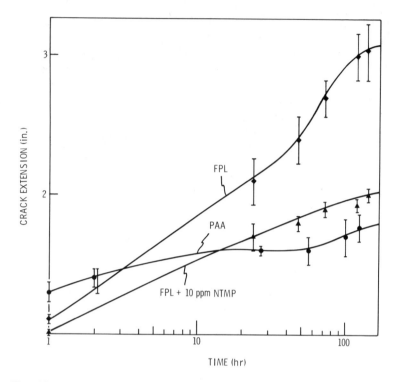

Fig. 12. *Crack extension vs time for FPL and PAA adherends and for FPL adherends treated with a 10 ppm NTMP solution.*[31]

the incubation time data, adsorption of the NTMP inhibitor always significantly improved the incubation time (for solution concentrations greater than 10 ppm).

Typical wedge-test results (Fig. 12) for FPL, FPL treated with a 10 ppm NTMP solution, and PAA surfaces demonstrated that the treatment of 2024 Al with 10 ppm NTMP significantly improved the durability of FPL-etched adherends. In fact, the performance of the NTMP-treated FPL adherend was nearly equivalent to that of PAA adherends.[31] Similar wedge tests of adherends with less than saturation coverage of NTMP, and with more than saturation coverage (obtained at elevated temperatures), indicated that long-term bond durability increased with coverage up to one monolayer.[31] Incubation time results also exhibited this trend, supporting the model that the oxide-to-hydroxide conversion process is responsible for the degradation of Al adhesive bonds in humid environments.

Therefore, one method of improving the long-term joint durability of FPL adherends is to treat them with a hydration inhibitor before bonding; another is to choose an improved surface treatment, such as phosphoric acid anodisation (PAA).

3.2. Phosphoric Acid Anodising Pretreatment for Aluminium Adherend Surfaces

The morphology of a phosphoric acid anodised (PAA) adherend,[27] shown in Fig. 13, differs from that of the FPL adherend (Fig. 4) in that: (i) the oxide produced by anodisation is considerably thicker; (ii) the hollow hexagonal cell structure, which exhibits a low profile in the FPL oxide, is much better developed in the PAA films; and (iii) the whisker-like protrusions are considerably longer on the PAA-treated surface.

Because of its more fully developed structure, it might be expected that if the adhesive or primer penetrated into the porous oxide, the PAA surface would provide better mechanical interlocking to a polymer and therefore exhibit a stronger bond than the FPL surface. Joint strength measurements comparing PAA and FPL surfaces[32] show increased durability while the increased degree of mechanical interlocking is shown in Fig. 14. In these studies, the 6061 Al panels were prepared using the standard PAA process and then coated with an adhesive-based primer applied according to specifications. The Al was then bent sharply until the primer cracked, allowing a cross-section of the oxide/primer interface to be observed with XSEM. The micrograph revealed that the primer penetrated completely into the porous oxide leaving absolutely no voids or empty regions, even at the bottom of the pores. Strong capillary forces were probably responsible for such a high degree of penetration, and the wettability of the oxide by the polymer material probably played an important role in achieving penetration. This mechanism assigns a somewhat more indirect role to the wettability factor than is ordinarily given to it and emphasises the kinetics of the process. Conventionally, good wettability is assumed necessary to achieve good bond strength because it implies strong inter-molecular bonding across the interface. This is undoubtedly a critical factor in the case when bonds are made to smooth surfaces. However, for porous surfaces the effect may be one of promoting

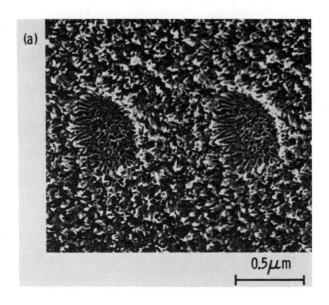

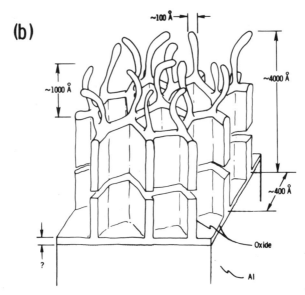

Fig. 13. (a) Ultra-high resolution stereo XSEM micrograph and (b) isometric drawing of the oxide morphology produced on an aluminium surface by PAA treatment (50 000×).[27]

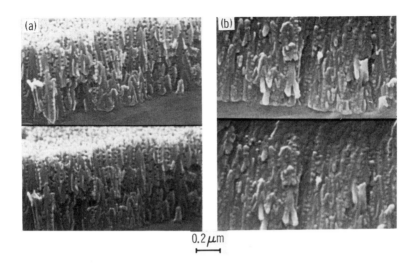

Fig. 14. Stereo micrograph of cross-section of PAA oxide on 6061 Al (a) before and (b) after being coated with adhesive-based primer.

penetration of the polymer to maximise the degree of interlocking and thereby achieve stronger joints.

Since the porous nature of the PAA oxide appears responsible for the success of this pretreatment process in promoting strong adhesive joints to Al, the kinetics of oxide growth and development of the whisker-like oxide morphology were investigated to learn how processing parameters affect the oxide morphology.[29] These results indicated that the development of the oxide structure proceeded in two stages as shown in the stereo micrographs of Fig. 15 and the thickness measurements of Fig. 16. The first stage (lasting ~3 min in a 18 wt% H_3PO_4 solution and longer in the standard 10 wt% solution), involved a fast linear growth during which the pore cell structure formed. The second stage was characterised by slower growth during which the fine whiskers were formed on top of the cells.

The discrepancy in the oxide thicknesses determined by XSEM and AES is a consequence of the porosity of the oxide film. Thicknesses derived from AES sputtering times are dependent on a calibrated sputter rate. This rate was determined with non-porous oxide films of known thickness. Because the time to sputter through a film is more a function of material removed than of actual physical thickness, it

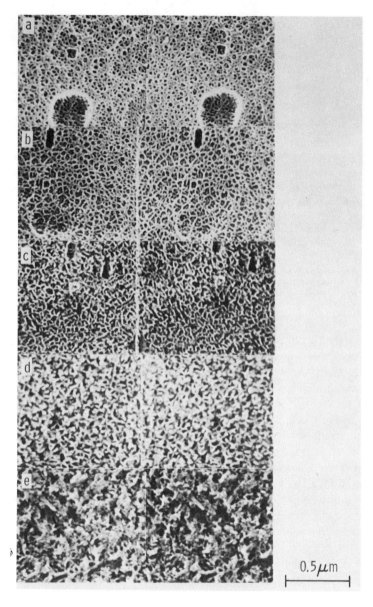

Fig. 15. *Stereo micrographs of PAA oxide morphology for 2024 Al in 10 wt%
phosphoric acid. Anodisation time: (a) 1 min, (b) 3 min, (c) 8 min, (d) 10 min,
(e) 20 min.*[29]

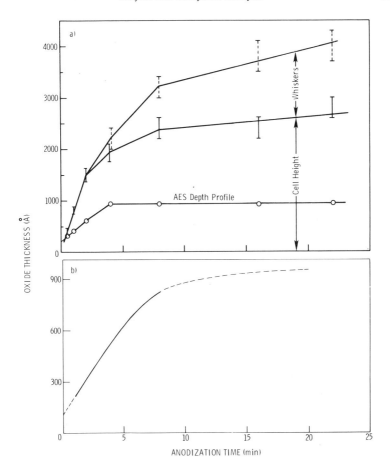

Fig. 16. PAA oxide thickness (for 2024 Al) as a function of anodisation time: (a) 18 wt% phosphoric acid determined by XSEM (vertical bars) and AES depth profiles (circles); and (b) 10 wt% phosphoric acid determined by AES depth profiles.[19,29]

provides an equivalent dense film thickness. Hence, the difference in thicknesses obtained by XSEM and AES is a measure of the porosity of the oxide. It therefore appears that during the second stage of growth, the total mass of oxide was constant, although its thickness (and also porosity) increased with time.

Failure of a PAA-prepared joint in a humid environment occurs, in analogy to the FPL oxide, when water (which can diffuse through the adhesive used) hydrates the Al_2O_3 and weakens it. The time required for this hydration, however, is much longer for PAA than for FPL and is comparable to that for NTMP-inhibited FPL. Evidently the PAA oxide inherently resists hydration.

To determine why the PAA oxide is hydration-inhibited and why this inhibition eventually fails, we used XPS to examine the surface after various exposures to humidity.[9] We found that the surface compositions could be expressed as linear combinations of Al_2O_3, $AlPO_4$ and H_2O. These compositions could then be plotted on to a surface behaviour diagram (Fig. 17) which was developed to trace the evolution of the surface composition during the hydration process. The

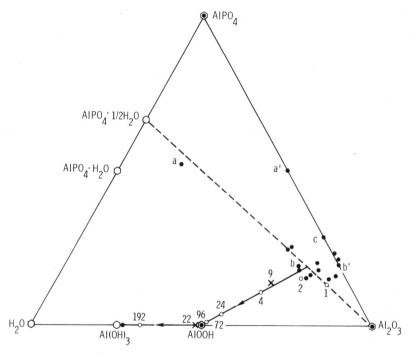

Fig. 17. The $AlPO_4$–Al_2O_3–H_2O surface behaviour diagram of the fresh PAA Al oxide surfaces (●). The open hexagons are theoretical compositions. All surfaces were rinsed in water after anodisation. Points a and a' and b and b' represent the same coupon before and after dehydration in the UHV chamber. Point c represents a coupon stored in a vacuum desiccator for 50 days. The numbers by some of the points denote the exposure (h) to 100% relative humidity at 50°C (○) or at 60°C (×).[9]

fresh PAA oxide surfaces and those of standards are represented by the solid dots.

Most of the data clustered at ~20% $AlPO_4$, which corresponds to approximately one monolayer coverage. Many of the data points lie off the $AlPO_4$–Al_2O_3 axis, indicating some adsorption of H_2O. This adsorption was dependent on the storage conditions of the sample and could be reversed by storing the samples in UHV for three days (a → a' and b → b'), or in a vacuum desiccator for 50 days (c). No morphological changes occurred as a result of this adsorption, and it was considered only a precursor to the hydration.

With exposure to high humidity (relative humidity 100% at 50°C or 60°C), the surface composition evolved directly to that of boehmite, AlOOH. Further hydration then occurred with the final hydration product being bayerite, $Al(OH)_3$. Corresponding stereo micrographs showed the hydration product growing around the whiskers, filling the pores, and finally completely covering the surface; X-ray diffraction identified the hydration products as boehmite and bayerite.

To help develop a model for the hydration process, we compared the experimentally determined results to expected surface composition changes for various possible reactions.[9] The results are shown in Fig. 18 and described below:

(a) *Adsorption of H_2O.* The surface layer of $AlPO_4$ on the fresh PAA oxide becomes $AlPO_4.nH_2O$. The $AlPO_4$ content does not change significantly, although the content of H_2O increases at the expense of detectable Al_2O_3. The surface composition thus evolves along the horizontal direction (path a in Fig. 18). The dehydration of $AlPO_4.nH_2O$ proceeds in the same way, but in the reverse direction.

This behaviour can also be adapted to a surface containing no phosphate. The surface Al_2O_3 becomes hydrated and less of the underlying Al_2O_3 is detected. The composition then follows the Al_2O_3–H_2O axis (path a' in Fig. 18).

(b) *Nucleation and growth of a hydrated phase with composition $Al_2O_3.mH_2O$.* The original surface of $AlPO_4.nH_2O$ remains unchanged, but becomes covered by an overlayer of $Al_2O_3.mH_2O$. Hence, the surface composition at any time is a linear combination of the original composition and that of the hydration product. Its path would be represented by a straight line on the surface behaviour diagram (path b in Fig. 18).

(c) *Dissolution of $AlPO_4$ or $AlPO_4.nH_2O$ without hydration of the underlying Al_2O_3.* The $AlPO_4$ and H_2O decrease proportionally while

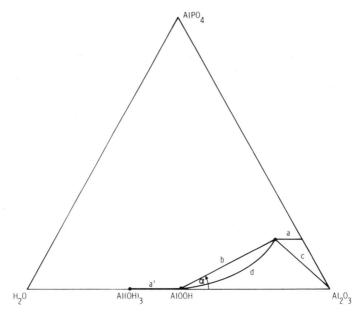

Fig. 18. *The surface behaviour diagram showing different evolution paths of surface composition: a, hydration of surface* $AlPO_4$ *to form* $AlPO_4.nH_2O$; *a′, hydration of surface* $Al_2O_3.nH_2O$ *to* $Al_2O_3.mH_2O$, $m > n$; *b, nucleation and growth of a hydrated-phase AlOOH; c, dissolution of* $AlPO_4.nH_2O$ *without hydration of underlying* Al_2O_3; *and d, dissolution of* $AlPO_4.mH_2O$ *followed by hydration of* Al_2O_3.

the underlying Al_2O_3 becomes more prominent. Again the evolution of the surface composition is along a straight line—the tie line between the original composition and the Al_2O_3 vertex (path c in Fig. 17).

(*d*) *Dissolution of* $AlPO_4.nH_2O$ *followed by hydration of the exposed* Al_2O_3. This final possible reaction can be considered a combination of reactions (*b*) and (*c*). The initial and final compositions are the same as (*b*), but the phosphate concentration initially decreases more rapidly in this example. The composition of the surface evolves along a curve bounded by tie line b and tie lines c and $Al_2O_3-H_2O$ (path d in Fig. 18). The deviation of this curve from the straight line is a function of the relative kinetics of the two mechanisms. In the limiting case where the dissolution of $AlPO_4$ proceeds slowly and it is the rate-limiting step, the hydration of the exposed surface follows quickly and the evolution of the surface oxide composition is very similar to that of case (*b*). These are, in fact, the expected relative rates of dissolution and hydration; the surface behaviour diagram alone cannot distinguish between mechanisms (*b*) and (*d*). To do so, additional measurements are needed to determine the amount and distribution of phosphate in the hydroxide overlayer.

In reaction (*b*) the phosphate does not dissolve and should be found below the hydration products; in reaction (*d*) the phosphate dissolves and very little should be found in the hydroxide layer. The corresponding Auger depth profiles of an original PAA oxide and of a hydrated PAA oxide are shown in Fig. 19. The fresh oxide shows a substantial amount of phosphorus detectable on the surface and in the oxide (on the surfaces of the cell walls). (The increase in the phosphorus signal before sputtering is due to an electron beam reduction of the phosphate to elemental phosphorus or to a phosphide and the corresponding Auger line shape change.[19] This change makes absolute Auger quantitative analysis difficult, but the relative differences which are of interest should be unaffected.) Very little phosphorus, on the other hand, is detected in the hydrated sample. Even with a five-fold increase in sensitivity, the phosphorus signal is just barely detectable above the zero level represented by Cu in the oxide. A small increase is observed at the interface, but the total amount of phosphorus present after hydration was estimated to be less than 20% of that originally present.

These results indicated that the dissolution of phosphate (case (*d*)) plays an important role in the hydration and can be explained by a three-step model of the hydration of PAA oxides.[9] The first step consists of the adsorption of water by the surface $AlPO_4$ layer. It involves no change in morphology and can be easily reversed by storing in a dry environment or by heating.

The second step comprises the slow dissolution of the inhibitive phosphate followed by the rapid hydration of the freshly exposed Al_2O_3 to boehmite, $AlOOH$. It is accompanied by a drastic change in morphology.

The third hydration step consists of the nucleation and growth of the bayerite phase, $Al(OH)_3$. Although XSEM micrographic and X-ray diffraction analysis indicate bayerite crystallites on top of the boehmite, the results are insufficient to determine whether the bayerite is converted from the boehmite via dissolution and redeposition or if it simply nucleates in the presence of the boehmite.

It is during the second stage that adhesive bonds fail in humid

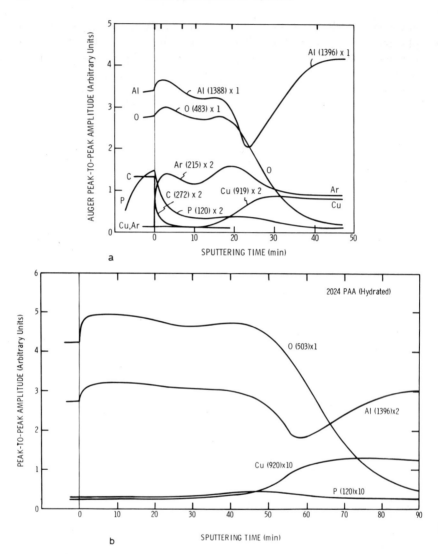

Fig. 19. (a) *Auger depth profile of unhydrated PAA oxide on 2024 Al. The numbers in parentheses are the kinetic energies of the Auger transitions used. The relative scale for each transition is given by the number following the parentheses. (b) Auger depth profile of PAA oxide on 2024 Al hydrated in 100% relative humidity at 50°C for 73 h. Note that the oxygen transition is different from that used for (a). The phosphorus signal in the oxide is barely detectable above the noise level given by the copper signal in the oxide and the phosphorus signal in the metal.*

environments. The analysis indicated that the slow dissolution of the phosphate is likely to be the rate-controlling step in the hydration and is responsible for the greater stability of PAA oxides relative to FPL oxides. This finding led to the investigation of monolayer hydration inhibitors for the FPL surfaces discussed earlier.

4. CONCLUSIONS

In this chapter we have attempted to provide an introduction to the capabilities and limitations of several surface analytical techniques used to study adhesive bonding. Because of space limitations, we have concentrated on those which we have found most useful for a variety of problems, but we do not intend to suggest that the others do not provide very useful, if not critical, information in other laboratories. For a more extensive discussion of each of these techniques, the reader is encouraged to consult the reviews listed in the bibliography.

We then discussed several investigations of different aspects of adhesive bonding to illustrate the applications of some of these techniques and the interplay among them. In many cases, a better understanding of the problem can be gained by exploiting the complementary information provided by the use of more than one technique.

REFERENCES

1. Wagner, C. D., Riggs, W. M., Davis, L. E., Moulder, J. F. and Muilenberg, G. E. *Handbook of X-ray Photoelectron Spectroscopy*, Perkin-Elmer, Eden Praire, Minn. (1979).
2. Carlson, T. A. *Photoelectron and Auger Spectroscopy*, Plenum, New York (1974).
3. Wagner, C. D. *Faraday Disc. Chem. Soc.*, **60** (1975), 291.
4. Wagner, C. D., Gale, L. H. and Raymond, R. H. *Anal. Chem.*, **51** (1979), 466.
5. Wagner, C. D., Zatko, D. A. and Raymond, R. H. *Anal. Chem.*, **52** (1980), 1445.
6. Wagner, C. D., Passoja, D. A., Hillery, H. G., Kinisky, T. G., Six, H. A., Jamsen, W. T. and Taylor, J. A. *J. Vac. Sci. Technol.*, **21** (1982), 933.
7. Wagner, C. D. *Anal. Chem.*, **44** (1972), 1050.
8. Carter, W. J., Schweitzer, G. K. and Carlson, T. A. *J. Electron Spectrosc. Relat. Phenom.*, **5** (1974), 827.
9. Davis, G. D., Sun, T. S., Ahearn, J. S. and Venables, J. D. *J. Mater. Sci.*, **17** (1982), 1807.

10. Davis, G. D., Buchner, S., Beck, W. A. and Byer, N. E. *Appl. Surf. Sci.*, **15** (1983), 238.
11. Davis, L. E., MacDonald, N. C., Palmberg, P. W., Riach, G. E. and Weber, R. E. *Handbook of Auger Electron Spectroscopy*, Perkin-Elmer, Eden Praire, Minn. (1976).
12. Coghlan, W. A. and Clausing, R. E. *Atomic Data*, **5** (1973), 317.
13. Madden, H. H. *J. Vac. Sci. Technol.*, **18** (1981), 677.
14. Ramaker, D. E. Auger spectroscopy as a probe of valence bonds and bands, in *Springer Series in Chemical Physics*, ISISS Proceedings, ed. R. Vanselow, Springer-Verlag, Heidelberg (1982).
15. Sun, T. S., Chen, J. M., Viswanadham, R. K. and Green, J. A. S. *Appl. Phys. Lett.*, **31** (1977), 580.
16. Solomon, J. S. and Baun, W. L. *Surf. Sci.*, **51** (1975), 228.
17. Davis, G. D., Natan, M. and Anderson, K. A. *Appl. Surf. Sci.*, **15** (1983), 321.
18. Wehner, G. K. The aspects of sputtering in surface analysis methods, in *Methods of Surface Analysis*, ed. A. W. Czanderna, Elsevier, Amsterdam (1975).
19. Sun, T. S., McNamara, D. K., Ahearn, J. S., Chen, J. M., Ditchek, B. and Venables, J. D. *Appl. Surf. Sci.*, **5** (1980), 406.
20. Holloway, P. H. and McGuire, G. E. *Thin Solid Films*, **53** (1978), 3.
21. White, H. W., Godwin, L. M. and Ellialtioglu, R. *J. Adhesion*, **13** (1981), 177.
22. Goldstein, J. I. Electron optics, in *Practical Scanning Electron Microscopy*, ed. J. I. Goldstein and H. Yahowitz, Plenum, New York (1975).
23. Goldstein, J. I. Electron beam–specimen interactions, in *Practical Scanning Electron Microscopy*, ed. J. I. Goldstein and H. Yahowitz, Plenum, New York (1975).
24. Newberg, D. E. and Yahowitz, H. Specimen preparation, special techniques, and application of the scanning electron microscope, in *Practical Scanning Electron Microscopy*, ed. J. I. Goldstein and H. Yahowitz, Plenum, New York (1975).
25. Aspens, D. E. *J. Vac. Sci. Technol.*, **18** (1981), 289.
26. Eichner, H. W. and Schowalter, W. E. *Report No. 1813*, Forest Products Laboratory, Madison, Wis. (1950).
27. Venables, J. D., McNamara, D. K., Chen, J. M., Sun, T. S. and Hopping, R. L. *Appl. Surf. Sci.*, **3** (1979), 88.
28. Sun, T. S., Chen, J. M., Venables, J. D. and Hopping, R. L. *Appl. Surf. Sci.*, **1** (1978), 202.
29. Ahearn, J. S., Sun, T. S., Froede, C., Venables, J. D. and Hopping, R. L. *SAMPE Q.*, **12** (1980), 39.
30. Venables, J. D., McNamara, D. K., Chen, J. M., Ditchek, B. M., Morgenthaler, T. I., Sun, T. S. and Hopping, R. L. *Proc. 12th Nat. SAMPE Symp.*, 909 (1980).
31. Ahearn, J. S., Davis, G. D., Sun, T. S. and Venables, J. D. Correlation of surface chemistry and durability of aluminium/polymer bonds, in *Adhesion Aspects of Polymer Coatings*, ed. K. L. Mittal, Plenum, New York (1983).

32. Kabayaski, G. S. and Donnelly, D. J. *Document* D641517, Boeing, Seattle (1974).
33. Chen, J. M., Sun, T. S., McNamara, D. K., Venables, J. D. and Hopping, R. L. *Proc. 24th Nat. SAMPE Symp.*, 1188 (1979).

BIBLIOGRAPHY

I. General Surface Science

Blakely, J. M. *Introduction to the Properties of Crystal Surfaces*, Pergamon, Oxford (1973).
Morrison, S. R. *The Chemical Physics of Surfaces*, Plenum, New York (1977).
Prutton, M. *Surface Physics*, Oxford University Press, London (1975).
Somorjai, G. A. *Principles of Surface Chemistry*, Prentice-Hall, Englewood Cliffs, N.J. (1972).
Tabor, D. *Contemp. Phys.*, **22** (1981), 215.

II. Surface Analytical Techniques

Archer, R. J. *J. Opt. Soc. Am.*, **52** (1962), 970. [Ellipsometry]
Aspnes, D. E. *J. Vac. Sci. Technol.*, **18** (1981), 289. [Ellipsometry]
Benninghover, A., *Surf. Sci.*, **53** (1975), 596. [SIMS]
Carlson, T. A. *Photoelectron and Auger Spectroscopy*, Plenum, New York (1974). [XPS, AES]
Chang, C. C. *J. Vac. Sci. Technol.*, **18** (1981), 276. [XPS, AES]
Chatterji, D. *The Theory of Auger Transitions*, Academic Press, New York (1976). [AES]
Czanderna, A. W. (ed.), *Methods of Surface Analysis*, Elsevier, New York (1975). [XPS, AES, SIMS]
D'Esposito, L. and Koenig, J. L. *Fourier Transform Infrared Spectroscopy*, Vol. 1, Academic Press, New York (1978). [FTIR]
Goldstein, J. I. and Yahowitz, H. (eds.), *Practical Scanning Electron Microscopy*, Plenum, New York (1975). [SEM]
Hansma, P. K. *Phys. Rep.*, **30C** (1977), 145. [IETS]
Hansma, P. K. (ed.), *Tunneling Spectroscopy: Capabilities, Applications, and New Techniques*, Plenum, New York (1981). [IETS]
Heinrick, K. F. J. and Newbury, D. E. *Secondary Ion Mass Spectrometry*, NBS Spec. Publ. 427 (1975). [SIMS]
Koenig, J. L. *Acc. Chem. Res.*, **14** (1981), 171. [FTIR]
McCrackin, F. L., Passaglia, E., Stromberg, R. R. and Steinber, H. S. *J. Res. Nat. Bur. Stand.*, **67A** (1963), 363. [Ellipsometry]
Powell, C. J. *Appl. Surf. Sci.*, **4** (1980), 492. [XPS, AES, SIMS]
Shott, P. N. and Field, B. O. *Surf. Interfac. Anal.*, **1** (1975), 63. [IETS]
Siegbahn, K., Nordling, C., Fahlman, A., Norberg, R., Hamrin, K., Hedman, J., Johansson, G., Bergmark, T., Karlsson, S. E., Lindgren, J. and Lindberg,

B. *Nova Acta Reg. Soc. Sci. Upsal. Serv. IV,* **20** (1967), Upsala; *Techn. Rept. AFML-TR-68-189,* Wright-Patterson AFB, Ohio, (Oct. 1968). [XPS]

Swingle, R. S. and Riggs, W. M. *CRC Critical Rev. Anal. Chem.,* **5** (1975), 262. [XPS]

Wagner, C. D., Riggs, W. M., Davis, L. E., Moulder, J. F. and Muilenberg, G. E. *Handbook of X-ray Photoelectron Spectroscopy,* Perkin-Elmer, Eden Praire, Minn. (1979). [XPS]

3

Kinetics and Mechanism of Environmental Attack

J. COMYN

Leicester Polytechnic, Leicester, UK

NOTATION

A	pre-exponential factor in the Arrhenius equation
$C, C_1, C_2, C_t, C_x, C_y, C_{xy}, C_\infty, c$	concentration
D	diffusion coefficient
D_0	pre-exponential factor in the Arrhenius equation for diffusion
E_a	activation energy for bond breaking
E_d	activation energy for diffusion
F	interionic force
F_x, F_y, F_z	diffusive flux
k	rate constant
$l = 2\ell$	film thickness, width or breadth of a joint
M_t	mass uptake at time t
M_∞	mass uptake at equilibrium
q_1, q_2	ionic charge
R	gas constant
r	interionic distance
S	swelling stress
T	absolute temperature
T_g	glass transition temperature
t	time
v^*	a constant
W_A, W_{AL}	work of adhesion
x, y, z	space coordinates
γ	surface free energy (subscript indicates phases)
γ^P	polar component of surface free energy
γ^D	dispersion component of surface free energy

ε_0 permittivity of a vacuum $= 8 \cdot 85 \times 10^{-12}$ F m^{-2}
κ relative permittivity (dielectric constant)
σ stress

ABBREVIATIONS

R.H.   relative humidity

Adhesives and Hardeners

AF 126-2	nitrile epoxide adhesive (3M)
BSL 312,	modified epoxide adhesives (Ciba–Geigy)
BSL 313A	
DAB	1,3-diaminobenzene
DAPEE	di(1-aminopropyl-3-ethoxy)ether
DDM	4,4′-diaminodiphenylmethane
DDS	4,4′-diaminodiphenylsulphone
DGEBA	diglycidyl ether of bisphenol A
DMP	tris(dimethylaminomethyl)phenol
FM 73	modified epoxide adhesive (American Cyanamid)
FM 123-5	nitrile epoxide adhesive (American Cyanamid)
FM 1000,	epoxide–nylon adhesive (American Cyanamid)
FM 1000 EP/15	
MNA	methyl nadic anhydride
Redux 775	vinyl phenolic adhesive (Ciba–Geigy)
TETA	triethylenetetramine
TGDDM	tetraglycidyl-4,4′-diaminodiphenylmethane

1.   INTRODUCTION

Water is the substance which gives the greatest problems in the environmental stability of adhesive joints. The great majority of bonded structures are exposed to moist air, and if the relative humidity is high then over a period of time the strength of joints usually declines. There have been many studies in the literature which have demonstrated this by exposing adhesive joints to either natural or laboratory climates. These include a study by Cotter[1] which involved exposing joints in aluminium or titanium alloy bonded with epoxide or phenolic-based adhesives to a hot/wet climate or a hot/dry climate in

Australia, or a temperate climate in England. Joints were also exposed in a humidity cabinet at 95–100% R.H. at 45°C, and hot/wet conditions were clearly revealed as the most damaging. Also, Brewis, Comyn and their co-workers[2-5] exposed joints in aluminium alloy bonded with a range of epoxide adhesives in a humidity cabinet at 95% R.H. at 50°C and found that in almost every case weakening took place. However, joints exposed to laboratory air which is, in fact, quite humid (~50% R.H. and 20°C) were unaffected. The typical behaviour of adhesive lap joints in moist air is illustrated by Fig. 1. It can be seen that at the higher humidity strength falls with time, but it tends to level out rather than fall to zero.

Small joints with overlaps of 12×25 mm or 25 mm $\times 25$ mm have been extensively used in durability studies and the time scale for their weakening is measured in months or years, the rate of weakening increasing with temperature.

There seem to be two fundamental problems which arise with water, its widespread occurrence and the fact that adhesives are hydrophilic. The polar groups which confer adhesive properties on a substance also make them hydrophilic. Further, the substrates themselves may be very hydrophilic.

Surface treatment of adherends has been seen as the best means of maintaining strength in the wet and there are many examples of this in the literature.[1,3,6,7] Materials which are based on epoxides, phenolics or acrylics are used as structural adhesives, and epoxides have been the most widely studied as regards their durability. Acrylic adhesives are

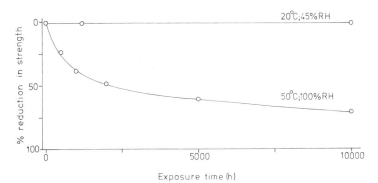

Fig. 1. Typical behaviour of the strength of lap joints on exposure to high and ambient humidity. Here the adhesive was FM 1000 with a nylon carrier and the adherend aluminium.[4]

relative newcomers,[8] but there is now a renewed interest in phenolic adhesives because of their apparently superior durability.[1]

Apart from their use in adhesives, epoxides are widely used as the matrix resin in fibrous composites and here again, water has a weakening effect. The effect of water on composites has been the subject of many papers in the literature some of which are referred to in the following text because of the many durability features shared by adhesive joints and composites. In a recent review Antoon and Koenig[9] have examined the moisture stability of glass-reinforced epoxide-composites.

What, therefore, is the mechanism by which water alters the strength of adhesive joints? It would seem that water may enter and alter a joint by one of a combination of the following processes, which have also been recognised[10] as possible contributors to the weakening effect which water has upon composites. Firstly, water may enter the joint by:

(a) Diffusion through the adhesive.
(b) Transport along the interface, a process which is often referred to as wicking. In this context the interface could be the adhesive/adherend interface, or if a fibrous carrier is incorporated in the adhesive then also the carrier/adhesive interface.
(c) Capillary action through cracks and crazes in the adhesive. This process is more likely to occur in joints which have been aged rather than in those that are freshly prepared.
(d) Diffusion through the adherend if it is permeable, such as might be the case with a composite.

Secondly, having entered a joint, water may cause weakening by one or a combination of the following actions:

(e) Altering the properties of the adhesive in a reversible manner. Plasticisation of the adhesive is such a process.
(f) Altering the properties of the adhesive in an irreversible manner, either by causing it to hydrolyse, to crack or to craze.
(g) Attacking the adhesive/adherend interface either by displacing the adhesive or by hydrating the metal or metal oxide surface of the adherend.
(h) Inducing swelling stresses in the adhesive joints.

These possibilities will now be examined in detail.

2. KINETICS OF WATER ENTRY

2.1. Prediction of Water Uptake from Diffusion Coefficient

If water enters a joint by diffusion, then it is possible to calculate the distribution of water using standard equations. These can be found in Crank's book on mathematics of diffusion[11] or in the book by Carslaw and Jaeger on the related phenomenon of heat conduction.[12]

The amount of substance diffusing in the x-direction across a plane of unit area in unit time is known as the flux, and is related to the concentration gradient $\partial c/\partial x$ by Fick's first law,

$$F_x = -D \; \partial c/\partial x \tag{1}$$

This can only be directly applied to steady-state diffusion, where concentrations at points within the system are not varying with time; this is clearly not the case when uptake is occurring. The build-up or decay of a diffusing species in a small-volume element is given by Fick's second law, which can be derived from the first law. The derivation considers that the change of concentration with time in the element is controlled by the fluxes crossing the six faces, as represented by Fig. 2. In Cartesian co-ordinates Fick's second law is:

$$\frac{\partial c}{\partial t} = D \left[\frac{\partial^2 c}{\partial x^2} + \frac{\partial^2 c}{\partial y^2} + \frac{\partial^2 c}{\partial z^2} \right] \tag{2}$$

If diffusion is restricted to the x-direction, such as is the case presented by a thin film absorbing a fluid, where diffusion into the edges of the film can be ignored, the second law can be simplified to

$$\frac{\partial c}{\partial t} = D \frac{\partial^2 c}{\partial x^2} \tag{3}$$

Alternative forms of the second law exist for non-Cartesian co-ordinate systems, and these may be of use in the cases of sorption by fibres or spheres.

Solutions are required to a form of the second law appropriate to the particular boundary conditions. The solution for the case of a semi-infinite film of thickness 2ℓ in an infinite bath is

$$\frac{C}{C_\infty} = 1 - \frac{4}{\pi} \sum_{n=0}^{\infty} \frac{(-1)^n}{(2n+1)} \exp\left[\frac{-D(2n+1)^2 \pi t}{4\ell^2} \right] \cos\left[\frac{(2n+1)\pi x}{2\ell} \right] \tag{4}$$

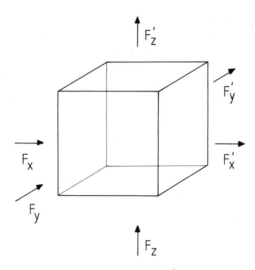

Fig. 2. Fluxes entering and leaving a Cartesian volume element.

and values of this equation are available in graphical form.[11,12] This film is initially free of diffusant and it is assumed that equilibrium between the bath and the surface of the film is established instantaneously on immersion. The faces of the film are located at $x = \pm \ell$.

Equation (4) gives the concentration of diffusant at points (x) within the film after various times. It may be integrated to give the total uptake by the film at time t; this is usually expressed as the fractional uptake M_t/M_∞.

$$\frac{M_t}{M_\infty} = 1 - \frac{8}{\pi^2} \sum_{n=0}^{\infty} \frac{1}{(2n+1)^2} \exp\left[\frac{-(2n+1)^2 \pi^2 Dt}{4\ell^2}\right] \qquad (5)$$

At short times eqn (5) approximates to eqn (6), so fractional uptake is linear with root time.

$$\frac{M_t}{M_\infty} = \frac{4}{l}\left(\frac{Dt}{\pi}\right)^{\frac{1}{2}} \qquad (6)$$

Shen and Springer[13] have shown that eqn (5) can be closely approximated by the simpler expression given below as eqn (7).

$$\frac{M_t}{M_\infty} = 1 - \exp\left[-7 \cdot 3 \left(\frac{Dt}{4\ell^2}\right)^{0.75}\right] \qquad (7)$$

If two films or slabs intersect at right angles, then their volume of intersection is a rectangular prism and the adhesive layer in a lap joint can be considered as a normal section of such a prism. These relationships are illustrated in Fig. 3. Equation (4) can be stated in the following form for the slab 1,

$$\frac{C_1}{C_\infty} = f(x, t, \ell_1) \tag{8}$$

and in a similar form for slab 2.

$$\frac{C_2}{C_\infty} = f(y, t, \ell_2) \tag{9}$$

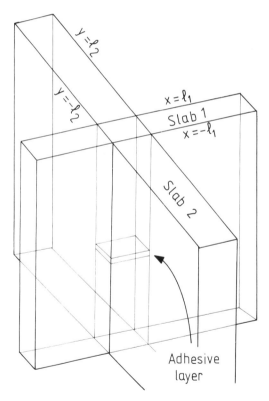

Fig. 3. Illustration that the adhesive layer in a lap joint is a section of a rectangular prism formed by the volume of intersection of two slabs at right angles.

Concentration at a point (x, y) within the prism or adhesive layer is then given by

$$\left[1 - \frac{C_{x,y}}{C_\infty}\right] = \left[1 - \frac{C_1}{C_\infty}\right]\left[1 - \frac{C_2}{C_\infty}\right] \tag{10}$$

The total amount of water taken up by the prism at time t is

$$M_t = \int_{-\ell_1}^{\ell_1} \int_{-\ell_2}^{\ell_2} C_1 C_2 \, dx \, dy \tag{11}$$

which can be written as

$$M_t = \int_{-\ell_1}^{\ell_1} C_1 \, dx \int_{-\ell_2}^{\ell_2} C_2 \, dy \tag{12}$$

The integrals have the form of eqn (5). In the case of a square adhesive layer the integrals are identical and so

$$M_t = \left[\int_{-\ell_1}^{\ell_1} C_1 \, dx\right]^2 \tag{13}$$

The use of these equations obviously requires reliable values of the diffusion coefficient, and we now examine how these can be obtained.

2.2. Measurement of Diffusion Coefficients

If a thin film of an adhesive is placed in liquid water or water vapour at a static pressure, then both the diffusion coefficient and solubility of water within the adhesive can be determined by following mass uptake of water as a function of time. An initial problem is to prepare thin films of even thickness and in the author's laboratories polytetrafluoroethylene, silicone rubber and tin have been found to be suitable substrates upon which adhesive films may be cast and subsequently removed. The duration of uptake experiments increases with the square of film thickness, and it is generally found that films of thickness 0·1–0·5 mm are most convenient. For uptake studies films should be semi-infinite (so that diffusion into the edges can be neglected) and placed in an infinite bath such as liquid water or vapour at constant relative humidity.

Brewis and Comyn[14] and their group in Leicester examined the liquid water uptake by a series of epoxide adhesives based on DGEBA. Film samples *ca* 0·35 mm thick were exposed to thermostatted distilled water in individual screw-capped jars at 25, 45 or 70°C,

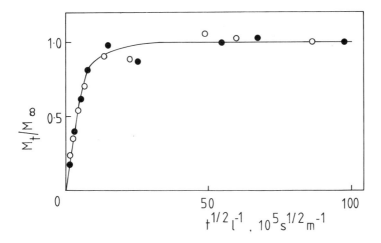

Fig. 4. Uptake of liquid water at 45°C by the DGEBA–DMP epoxides. Results of two experiments are shown.[14]

and in some cases exposure was to water at 100°C in a Soxhlet apparatus. Films were periodically removed for careful surface drying and weighing. Masses from such experiments can be plotted in the form of fractional uptake (M_t/M_∞) against root time in accordance with eqn (6) and all plots in this study showed an initial linear region of uptake up to $M_t/M_\infty > 0.6$. Such graphs are often referred to as uptake or sorption plots and diffusion coefficients can be evaluated from the slope of the linear region.

Recent work has shown that changes in length on water swelling can be used as an alternative to changes in mass in determining diffusion coefficients by eqn (6).[15,16]

An example of a mass uptake plot for an epoxide in liquid water at 45°C is shown in Fig. 4. Linear uptake of this type is termed Fickian and is typical of diffusion in a polymer in the leathery state (i.e. above its glass transition temperature) and not therefore what might be expected for diffusion in glassy epoxides. In general, glassy polymers show non-Fickian S-shaped uptake plots. Fickian and non-Fickian sorption have been compared by Fujita[17] and characteristics of Fickian diffusion which are of relevance to water uptake by adhesive films are as follows:

(i)   For both absorption and desorption plots of M_t/M_∞ against $t^{1/2}$

are initially linear, and this linearity extends to at least $M_t/M_\infty = 0{\cdot}6$ for absorption.

(ii)   Above the linear region, curves are concave against the abscissa.

(iii)   Plots of M_t/M_∞ against $t^{1/2}/l$, which are termed reduced sorption curves, should coincide for films of different thickness.

(iv)   Plots of M_t/M_∞ against $t^{1/2}$ for sorption and desorption will only coincide when D does not vary with concentration of the sorbant.

In recent years quite a large number of uptake plots for water in adhesives or in composites have appeared in the literature, and very frequently the characteristics of Fickian sorption as described in (i) and (ii) above have been evident, even though the glass transition temperature was in many cases considerably above the experimental temperature, so making non-Fickian diffusion likely. Glass transition temperatures, T_g, for the epoxide–Novolac resin Narmco 5208 have been reported at 250°C[18] and 216°C,[19] but several Fickian uptake plots have been reported both for the neat resin and for some carbon fibre composites with this as matrix resin, at temperatures considerably below quoted T_g, including ambient.[10,20–22] Similar cases include Fickian water uptake at 23°C by composites and resins based on TGDDM–DDS which has a glass transition temperature of 250°C,[18] and uptake at ambient temperature by composites with a matrix based on TGDDM–DDS and boron trifluoride monoethylamine and having a glass transition temperature of 178°C.[23] However, an exception to Fickian uptake has been reported for a resin formed from TGDDM and dicyandiamide.[24]

Epoxides cured with aliphatic amines have lower values of T_g than those cured with aromatic amines. DGEBA cured with TETA[25] and DAPEE[14,26] show Fickian water uptake. The epoxide–polyamide adhesive FM 1000 has a T_g when dry of 34°C and upon equilibriation with liquid water at room temperature its T_g is depressed to −32°C. Nevertheless, water uptake experiments at 25°C on this adhesive, during which the material changes from its glassy to leathery state, still produce Fickian uptake curves.[2]

Manson and Chiu[27] measured the water permeability of some films based on DGEBA, phenyl glycidyl ether and the polyamide Versamid 140, over a range of temperatures covering T_g. They found that plots of log (permeability) against reciprocal temperature consisted of two straight-line regions which intersected at about T_g, and that the activation energy was greater for the leathery than for the glassy state. Such a relationship between activation energies for the diffusion of helium, neon, argon, oxygen and hydrogen in high molar mass poly(vinyl acetate) was reported by Meares,[28] who accounted for the lower activation energy in the glassy state by there being a smaller zone of activation than in the leathery state.

Should samples of the same adhesive film be exposed to liquid water and its saturated vapour in equilibrium, then the zeroth low of thermodynamics requires their equilibrium uptakes to be the same. Equations 4, 5 and 6 assume that saturation of the infinitesimal surface layer is instantaneous and hence identical rates of uptake should be recorded for the liquid and vapour exposed films; in fact identical uptake plots should be seen. Liquid and vapour uptake observations on the adhesive DGEBA–DAPEE,[14,26] on FM 1000[2] and on some carbon fibre composites[29] give support to this view. However, Antoon and Koenig[9] note that with composites moisture uptake from liquid water is generally greater than that from saturated air. This difference could be due to some capillary attraction for liquid water or simply to the difficulty in obtaining truly saturated vapours. The difficulty has been used to explain this apparent violation of the zeroth law in the case of rubbers immersed in solvent or hung in the vapour.[30]

If an uptake experiment is undertaken from a single infinite bath, (e.g. liquid water) with which the epoxide establishes an equilibrium uptake M_∞, then the diffusion coefficient obtained from the plot of M_t/M_∞ against $t^{\frac{1}{2}}$ is an average over the concentration range 0 to M_∞. On account of the large fraction of free volume associated with liquids, they generally plasticise polymers. Because of this it might be anticipated that D might depend quite strongly on M_∞, and this is the situation which has been observed for many polymer–liquid combinations.[17]

The dependence of D upon concentration can be evaluated by performing uptake experiments from vapours at different pressures. This was done by Comyn *et al.*[26] on the adhesive based on DGEBA–DAPEE at 48°C. Films of adhesive were suspended in water vapour from the arm of a recording microbalance. In all cases uptake was found to be Fickian, and although their results show some experimental scatter, they clearly indicated that D is independent of concentration in this particular system. Further evidence for this was provided by the identical values of D obtained from sorption and desorption experiments. The constancy of D over a concentration range, although it may be unusual, greatly simplifies the calculation of water distribution in joints. It has been observed for some other water–epoxide systems,[31] but there are reports of systems where D increases with water concentration.[32,33] Daniely and Long[15] showed that water uptake increases with the extent of cure for a TGDDM–DDS adhesive, on account of the hydroxyl groups produced on curing.

Comyn et al.[26] also reported the dependence of equilibrium uptake upon vapour pressure for the DGEBA–DAPEE resin. Their results, which are shown in Fig. 5, are similar to a BET type III isotherm. They pointed out that a possible interpretation of the constant D and the shape of the uptake isotherm is that most of the water exists in clusters within the epoxide. A relatively small amount of water remains molecularly dispersed in the adhesive and, except at very low overall concentrations, its amount remains constant over the whole range of vapour pressure. Only the dispersed water molecules will contribute to the diffusion process. A similar isotherm has been reported for an epoxide composite[10] and a BET type II isotherm has been observed for water in a resin formed from TGDDM–DDS,[34] whilst a linear isotherm passing through the origin was reported[20] for composites based on Narmco 5208.

Whilst many adhesives do attain an equilibrium uptake with water as demonstrated by a level plateau region in uptake plots, this is not always the case. Clearly the lack of equilibrium in film uptake experi-

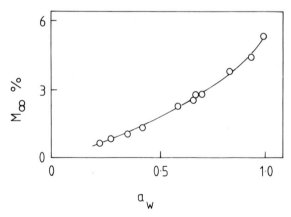

Fig. 5. Dependence of water solubility in the epoxide DGEBA–DAPEE at 48°C upon vapour activity.[26]

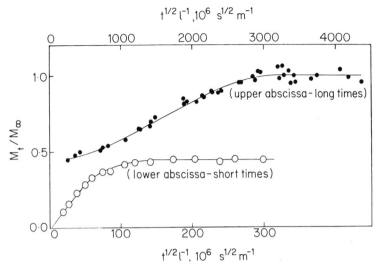

Fig. 6. *Uptake plot for the epoxide DGEBA–DDM from water at 100°C.*[14]

ments is a source of doubt in calculating D and also in evaluating saturation levels in adhesive joints.

Some adhesives show a second uptake stage. An example is the DGEBA–DDM resin in water at 100°C, for which the uptake plot is shown in Fig. 6; at 25, 45 and 70°C the same resin showed a single Fickian uptake stage.[14] This tendency to undergo a second uptake process at higher temperatures was noted for the adhesive FM 73 by Althof,[35] who observed that at 70°C and 95% R.H. a distinct second sorption stage appeared, but at lower temperatures (20, 40 and 50°C) at 95% R.H. this was absent.

Other adhesives show weight losses after the initial uptake. One example is the DGEBA–boron trifluoride monoethylamine resin at 70°C in water, which is illustrated in Fig. 7. At 25°C this resin gave an equilibrium plateau, but weight loss persisted at some intermediate temperatures.[14] The uptake plot from liquid water at 50°C by the adhesive FM 1000 shows weight loss, but uptake from the saturated vapour at the same temperature gives a Fickian uptake plot with a plateau. In the liquid, weight loss is due to leaching out of some component which is not lost to the vapour phase. This is completely removed in the first liquid sorption process so that re-exposure after drying gave an uptake plot with a plateau.

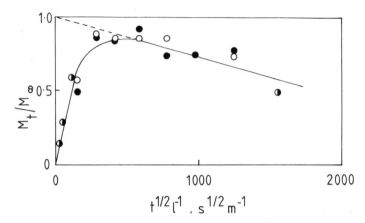

Fig. 7. Uptake of liquid water at 70°C by the DGEBA–boron trifluoride
monoethylamine. Results of two experiments are shown.[14]

Knowledge of the mechanisms of these secondary sorption or desorption processes is slight, but speculation on them is permissible.
Leaching out, which has already been noted, may be of an unreacted
component in the adhesive or of a product of hydrolysis. Another
possibility is the further crosslinking in the water-plasticised adhesive
at high temperature. Cracking or crazing of the adhesive is a possible
reason for increased uptake that is discussed later.

2.3. Comparison of Predicted and Measured Water Uptake

Althof[36,37] made some aluminium/adhesive/aluminium sandwiches
with the modified epoxide adhesives BSL 313A and FM 73, the epoxide nitrile adhesives AF 126-2 and FM 123-5, and the vinyl–phenolic
adhesive Redux 775. It should be noted that FM 73, AF 126-2 and
FM 123-5 contained fibrous carriers designed to assist in handling of
the adhesives and in controlling glue-line thickness. The sandwiches
were 100 mm in length and 5, 10, 20 or 30 mm in width; their
uptake from unsaturated water vapour was followed by weighing and
values of D and M_∞ were obtained which compared reasonably well
with values from film uptake experiments. Values of D are shown in
Table 1. In some cases opening of the sandwich edges was observed,
and in others corrosion at the edges took place. Both these effects
could contribute to the significant differences between diffusion coefficients noted in some instances.

TABLE 1

COMPARISON OF DIFFUSION COEFFICIENTS MEASURED FROM MASS UPTAKE
BY ADHESIVE FILMS (f), AND SANDWICHES (s) OF ADHESIVE
BETWEEN ALUMINIUM. AFTER ALTHOF[36]

Temp. (°C)	Rel. humidity (%)	$D/10^{-13} m^2 s^{-1}$ Adhesive							
		FM73		BSL313A		AF126-2		FM123-5	
		f	s	f	s	f	s	f	s
20	70	2·8	3·6	1·4	1·7	1·7	2·2	—	—
20	95	1·7	3·9	0·7	1·8	—	—	—	—
40	70	8·0	10·3	1·9	3·9	5·0	6·1	—	—
40	95	4·2	4·1	0·9	28	—	—	—	—
50	95	1·2	1·5	1·1	5·0	—	—	1·4	7·5
70	95	—	—	0·8	3·6	—	—	—	—

Further, Althof took sandwiches measuring 30 mm × 100 mm and exposed them to water vapour. After removing and discarding 15 mm from each end, the remainder was cut into strips measuring 3 mm × 70 mm. Water levels in the strips were determined by weighing before and after careful drying. Figure 8 compares some measured water distributions in sandwiches with those calculated from a diffusion coefficient obtained by film uptake experiments.

Brewis and Comyn and their co-workers[38] made sandwiches of adhesives between aluminium alloy measuring 25 × 12·5 mm. The adhesives used were the epoxide–polyamide adhesive FM 1000 and the modified epoxide adhesive BSL 312, as well as versions of both these adhesives containing knitted nylon carriers. The sandwiches were exposed to isotopically labelled water vapour, ^3HOH, at 50°C for periods up to six weeks, after which they were cut into 25 identically shaped rectangles, as shown in Fig. 9. Here the value shown for each piece is its fractional uptake measured by liquid scintillation counting after swelling the piece in dioxan. Because of the symmetry of the sandwiches, some pieces are identical with others. Identity is indicated by the letters A to I in Fig. 9. Measured uptakes are compared with those calculated from eqn 10 in Tables 2 and 3; individual values of C_t/C_∞ were calculated at several points within each piece and an average taken. There are three entries for each piece. The upper one is calculated from the diffusion coefficient and the middle and lower ones were measured for the uncarried adhesives respectively.

Moisture content, %
4
3
2
1
0
15
30
Distance, mm

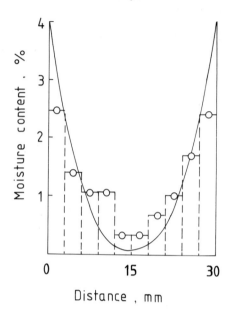

Fig. 8. *Moisture distribution across sandwiches bonded with the adhesive FM 73 after 60 days exposure to air at 70°C and 95% R.H. Experimental points are compared with a line calculated using eqn (9). After Althof.*[37]

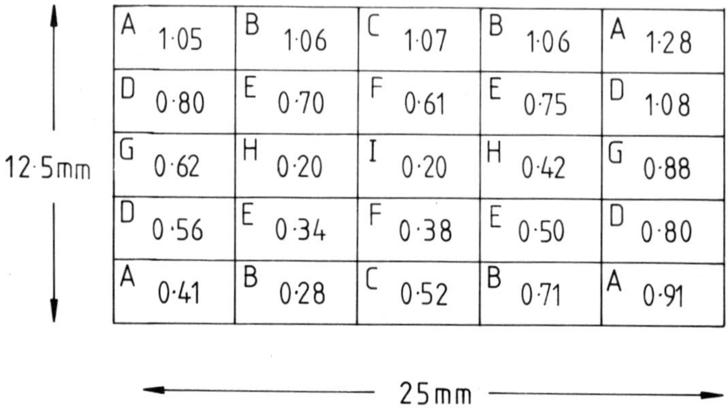

A 1·05
B 1·06
C 1·07
B 1·06
A 1·28
D 0·80
E 0·70
F 0·61
E 0·75
D 1·08
G 0·62
H 0·20
I 0·20
H 0·42
G 0·88
D 0·56
E 0·34
F 0·38
E 0·50
D 0·80
A 0·41
B 0·28
C 0·52
B 0·71
A 0·91
12·5mm
25mm

Fig. 9. *Fractional uptakes of isotopically labelled water in segments of an aluminium alloy FM 1000 sandwich after exposure to vapour for 100 h at 50°C.*[38]

TABLE 2

FRACTIONAL UPTAKE BY PIECES BONDED WITH FM 1000 AND FM 1000/EP 15 AFTER SIX WEEKS EXPOSURE TO LABELLED WATER VAPOUR[38]

A	0·93	B	0·83	C	0·79
	0·91±0·37		0·77±0·36		0·79
	0·86±0·19		0·70±0·13		0·82
D	0·85	E	0·61	F	0·51
	0·81±0·21		0·57±0·19		0·49
	0·85±0·06		0·39±0·10		0·32
G	0·83	H	0·55	I	0·45
	0·75		0·36		0·20
	0·66		0·09		0·02

Overall mean values for joint 0·75 (calculated)
0·69 (measured—no carrier)
0·60 (measured—carrier)

With FM 1000, uptake in the outer pieces was the same for both carried and uncarried adhesives, and these values agreed with the calculated values. In the inner pieces measured water was significantly less than predicted levels, and this was particularly the case when carrier was present. Overall mean levels of uptake by the sandwiches were in quite good agreement.

TABLE 3

FRACTIONAL UPTAKE BY PIECES BONDED WITH BSL312 AND BSL312/5 AFTER SIX WEEKS EXPOSURE TO LABELLED WATER VAPOUR[38]

A	0·87	B	0·68	C	0·61
	0·57±0·06		0·31±0·06		0·32
	0·99±0·23		0·55±0·07		0·53
D	0·75	E	0·37	F	0·23
	0·39±0·08		0·06±0·01		0·08
	0·61±0·08		0·095±0·006		0·21
G	0·74	H	0·27	I	0·18
	0·44		0·02		0·04
	0·60		0·045		0·07

Overall mean values for joint 0·58 (calculated)
0·28 (measured—no carrier)
0·47 (measured—carrier)

With BSL 312, the carried adhesive had uptake levels which again were comparable with the calculated values in the outer pieces, but uptake with the uncarried adhesive was distinctly lower. Towards the centre of the joint, measured water levels fell below calculated ones and again levels for uncarried adhesives were lowest.

It would seem from these results that the carrier in the FM 1000 adhesive hinders the passage of water. The capacity of a material to transmit a permeant is indicated by its permeability coefficient, which is the product of diffusion coefficient and solubility coefficient; the product DC_∞ was used as an approximate estimate of this quantity, and it is compared for the adhesives and carriers in Table 4. It illustrates that the FM 1000/EP 15 combination represents a relatively impermeable carrier in a permeable matrix, and the carrier so provides an obstacle to water diffusing through the adhesive, which is in accordance with the observed results. In the BSL 312/5 system, however, carrier and adhesive have much closer permeabilities, and would be expected to have similar efficiencies in transporting water into joints. This is again in accord with the results presented.

Support for the relative inertness of carriers in the process of water entry into joints comes from a study of their effect on the strength of double lap joints in aluminium alloy, exposed to saturated air at 50°C.[2,5] Adhesives based on BSL 312 and FM 1000 were used in this study, and joints were exposed for up to 10 000 h. It was found that carriers did not alter the durability characteristics of unstressed joints.

Schulte and DeIasi[39] exposed sandwiches of a carbon fibre epoxide composite bonded with FM 300 epoxide adhesive to a ^2H_2O environment at 70% R.H. and 77°C for various exposure times and then sectioned them by cutting with a diamond wheel at liquid nitrogen temperatures. Deuterium concentration at points on the cut faces could be measured by bombarding them with the $^3He^+$ ions from a

TABLE 4

DIFFUSION COEFFICIENTS AND EQUILIBRIUM WATER UPTAKE AT 50°C[38]

Substrate	$D/10^{-13} m^2 s^{-1}$	C_∞ (g per 100 g)	DC_∞
FM 1000	34 ± 6	13·4	460
BSL 312	18 ± 5	2·2	40
Carrier EP 15	$4·4 \pm 0·9$	7·2	32
Carrier from BSL 312/5	$2·9 \pm 0·6$	7·3	21

Van de Graaff accelerator and measuring the protons produced by the following reaction:

$$^2H + {}^3He^+ \rightarrow H^+ + {}^4He$$

The sandwiches measured approximately $25\,mm \times 12 \cdot 5\,mm$ and were *ca* $1 \cdot 9\,mm$ thick, with a thin glue line. Diffusion coefficients of ^2H_2O in composite and adhesive measured by mass uptake were $3 \cdot 7 \times 10^{-13}\,m^2s^{-1}$ and $1 \cdot 3 \times 10^{-12}\,m^2s^{-1}$ respectively. Measured moisture content along the glue line was compared with that calculated on the basis of one-dimensional diffusion along the glue line. Some results are shown in Figs 10 and 11. In all cases, measured water is greater than the calculated value and it can be seen that after 6 or 12 days exposure all water which has entered the central region of the adhesive must have done so by diffusion through the composite as one-dimensional diffusion predicts a zero water level, the composite presenting a shorter path to the centre of the joint.

The strengths of single lap joints in carbon fibre composite have been measured after exposure to 97–98% R.H. at 45°C by Parker.[40] Three different composites were used with two different adhesives and with a relatively impermeable adhesive the time needed for joint strength loss was similar to the time needed for the composite to reach

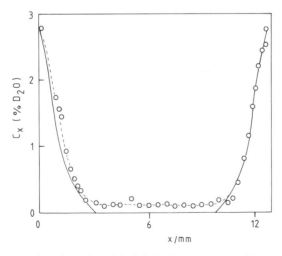

Fig. 10. *Measured and predicted (solid line) moisture profile for the adhesive layer in a composite lap joint after exposure to 70% R.H. at 77°C for 6 days.*[39]

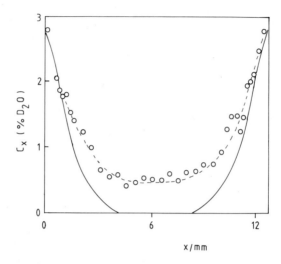

Fig. 11. Measured and predicted moisture profile for the adhesive layer in a composite lap joint after exposure to 70% R.H. at 77°C for 12 days. The solid line is calculated with the assumption that diffusion occurs only through the adhesive.[39]

saturation; loss of strength was mainly attributed to a decline in composite properties.

The durability of joints with glass-fibre reinforced polyester adherends bonded with some rubber-modified acrylic adhesives has been examined by Bowditch and Stannard.[41] Lap joints exposed to 90% R.H. at 70°C at first showed a high rate of strength loss but strength tended to level out after about 250 days of exposure. At short exposure times failure was generally within the GRP substrates but this was replaced by failure which was predominantly adhesive at inter-mediate times. At long times, however, there was a return of substrate failure and it was proposed that these changes in the type of failure were due to the different rates at which adhesive and polyester matrix degrade in the experimental conditions. Measurement of strength changes of GRP laminates with exposure (they lost about 38% of their strength over 250 h) supported this view. Similar experiments under the less stringent conditions of immersion in water at 40°C for up to 150 days simply showed decreasing substrate failure with time.

These various studies demonstrate that water distribution in joints with impermeable adherends can be accounted for by assuming Fickian diffusion in the adhesive layer, and where one or more adherends is permeable this provides a further path for water to diffuse into the joint. It also seems clear that carrier fibres do not generally have much effect on the rate of water entry. The contribution of carriers to the rate of ingress can be accounted for by diffusion along the fibres, so making it seem that water is not transported along the fibre/adhesive interface.

However, these studies employed carefully prepared substrates whereas it is in joints with less rigorous surface treatments that transport along the substrate/adhesive interface would seem more likely. Certainly, on the basis of the studies so far reported, interfacial transport cannot be excluded as a possible mechanism of water entry into adhesive joints.

Transport along the interface may be important when a silane primer has been applied to the substrate (I am grateful to Dr W. C. Wake for pointing this out to me). Silanes have been used for some time to treat glass fibres in the hope of improving the durability of GRP composites and it has been suggested that water diffusion along the glass/resin interface may be at least 450 times as fast as through the resin.[42] Kadotani[43] has claimed that an abnormal resin layer exists at the fibre/matrix interface in a glass–epoxide composite, and that this layer has a high water absorption and hence would be more permeable than normal. Evidence for this was based on a dielectric study of composites prepared from silane treated E-glass in an acid anhydride cured epoxide matrix to which 0·4% of water had been added before curing. Some samples were immersed in boiling water for up to 24 h. Water causes some hydrolysis, and the products of this would appear to be preferentially adsorbed at the fibre surface.

Silanes have been used to pretreat metal surfaces prior to adhesive bonding with consequent improvements in durability[44–47] and it has also been demonstrated that they produce polysiloxane coatings on iron,[46,47] steel,[44,45,48] aluminium[46] and aluminium oxide in the form of sapphire.[49,50] Bascom[48] showed that part of the coatings could be removed from steel by rinsing in acetone but there always remained regions of more firmly held material. It may well be the case that silanes lead to more durable interfaces but they may greatly assist the entry of water into adhesive joints. This is because polysiloxanes are about the most permeable polymers known. A detailed study by Barrie and Machin[51] examined water transport in three polydimethylsiloxanes; the diffusion coefficient of water in a Midland Silicone product seems typical with $D = 3 \cdot 5 \times 10^{-9}$ m²s⁻¹ at 26·3°C whereas a typical value for water in an epoxide at this temperature would be $1 \cdot 5 \times 10^{-13}$ m²s⁻¹,[14] so making it seem that water diffusion in the interfacial primer layer would be about 20 000 times faster than in the adhesive.

2.4. Dependence of Joint Strength on Water Content

Comyn and Brewis and their co-workers have observed a linear dependence of joint strength upon total water uptake for a number of adhesively bonded aluminium alloy lap joints. Some results are shown in Figs 12 and 13 for the adhesive DGEBA–DAPEE,[3] which has been demonstrated to have a diffusion coefficient independent of concentration, so adding confidence to the calculated water uptakes. It can be seen that the linearity persists to high levels of water uptake. The joints in this study were $12 \cdot 5 \times 12 \cdot 5$ mm single lap joints and exposure for about one year was required to achieve the high levels of uptake. The results in Figs 12 and 13 are for five different metal surface pretreatments, and it can be seen that even though one group of

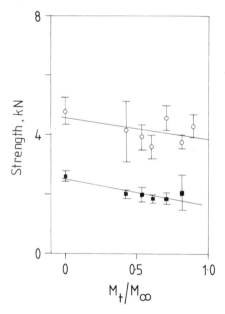

Fig. 12. *The linear relationship between joint strength and fractional water uptake for aluminium alloy lap joints bonded with DGEBA–DAPEE. Surface treatments*: O, *anodising method II*; ■, *by sandblasting.*[3]

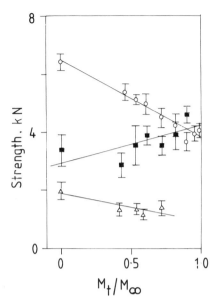

Fig. 13. *The linear relationship between joint strength and fractional water uptake for aluminium alloy lap joints bonded with DGEBA–DAPEE. Surface treatments to the alloy were by etching in chromic acid* ○, *anodising in chromic acid by method I* ■ *and degreasing in solvent* △.[3]

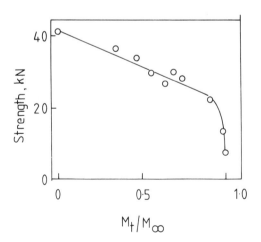

Fig. 14. *Dependence of joint strength upon water uptake for aluminium alloy double lap joints bonded with FM 1000.*[4]

anodised joints became stronger on exposure the linear dependence was still observed. A possible reason for this strengthening is relief of stresses within the joints by water plasticisation. Degreased and sandblasted joints are relatively weak when dry, but otherwise are not apparently any more susceptible to exposure than etched or anodised joints. A similar linear relationship was observed for joints using the modified epoxide adhesive BSL 312 with chromic acid etched aluminium alloy adherends,[5] but in the case of the epoxide–polyamide adhesive FM 1000[2,4] a marked departure from linearity occurred at high uptake. This departure, shown in Fig. 14, probably represents the attack of water on the interface after longer exposure.

3. EFFECT OF WATER ON THE ADHESIVE

Having examined the rate of water entry into adhesive joints, it remains to consider the mechanism whereby it causes weakening. Water could influence the adhesive or the interface and in the case of permeable adherends changes in substrate properties could also occur. Some of the processes which might occur would be reversible and so accompanied by recovery in strength on removal of water but others would be irreversible and the strength would be irrecoverable.

3.1. Plasticisation and Swelling
Plasticisation and swelling are both reversible processes. Water depresses the glass transition temperature of adhesives and lowers modulus and strength.

Several attempts have been made to relate the depression of T_g of adhesives by water to free volume theories of the glass transition using the equations of Kelley and Beuche[52] and Fox.[53] The application of such equations requires a value for the glass transition temperature of water and this has been reported in the range -134 to $-138°C$.[54–57] Using a T_g (H_2O) in this region, Morgan and Mones[58] applied the Kelley–Beuche theory to some experimental data of Browning on the TGDDM–DDS–water system and found that the actual glass transition temperatures of wet resins were well below predicted values. However Browning[23,59,60] found very good agreement between measured and predicted values of T_g if a value of $+4°C$ was used for the glass transition temperature of water.

Water is an unusual fluid, and is more complex in structure and behaviour than the organic plasticisers used to modify the properties of commercial plastics. It may well be that equating its T_g parameter to the temperature of maximum density is a better measure of its free volume contribution to a polymer–water mixture. Brewis, Comyn *et al.*[14] analysed the T_g depression of a range of epoxides using the Fox equation. All T_g values were depressed on first absorbing water (Table 5) and agreement between the predicted and experimental values was probably within experimental error when the following hardeners were used: DAPEE, DAB, DMP and boron trifluoride monoethylamine. In other cases there was less plasticisation than predicted, and this could be interpreted by water being clustered within the adhesive rather than being completely dispersed. An item of further interest which is recorded in Table 5 is that on prolonged exposure to water, the T_g of each adhesive studied increased, and in two cases it increased to a temperature above the T_g of the dry adhesive. Thus it would seem unlikely that plasticisation can satisfactorily explain the long term durability of joints, although doubtless it is a contributory factor in some cases. One such case is illustrated by some results using FM 1000 adhesive[61] which are shown in Fig. 15. FM 1000 is particularly susceptible to water plasticisation and the Figure shows strengths of some aluminium alloy single lap joints, $12 \cdot 5 \times 12 \cdot 5$ mm, at various temperatures; dry joints are compared with some made wet by exposure at 100% R.H. at 50°C for 1000 h, and calculated to have a fractional water content of $0 \cdot 86$. The number shown against each point is the percentage of interfacial failure. Wet and dry joints exhibit similar

TABLE 5

GLASS TRANSITION TEMPERATURES OF WET AND DRY ADHESIVES FORMED FROM DGEBA WITH VARIOUS HARDENERS[14]

Hardener	Measured T_g (°C)			T_g (°C) from Fox eqn
	Dry	After equilibration	After 10 months in water	After equilibration
DAPEE	67	37	49	44
TETA	99	86	111	76
DAB	161	143	157	139
DDM	119	110	130	92
DMP	68	51	54	47
$BF_3.C_2H_5NH_2$	173	155		151

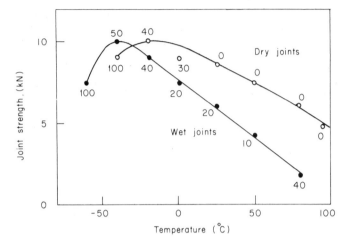

Fig. 15. *The strength of wet and dry lap joints with FM 1000 adhesive against temperature*: ○, *dry joints*; ●, *joints preconditioned for 1000 h at 50°C and 100% R.H.*[61]

strength–temperature curves but the wet curve is shifted to lower temperatures by about 30–50°C, an amount similar to the water-induced depression of T_g. Hence in this case the depression of T_g can be used as a shift factor to relate the strength–temperature curve for the dry adhesive to the wet one.

Water entering a joint may cause swelling and thereby introduce stresses into that joint which could weaken it. There are several studies in the literature demonstrating dimensional changes of adhesive materials on exposure to water including those of Gazit[62] and Shirrell and Halpin.[10] The latter showed examples of neat resins and composites where swelling strain was proportional to moisture content, and as has been noted earlier, these effects can be used to measure D.

Weitsman[63] has undertaken an analysis, based on variational principles, of the stresses introduced into adhesive joints by water swelling caused by Fickian diffusion. The swelling induced in the x-direction is given at relatively short times by eqn (14).

$$S(x) = S_0 \operatorname{erfc} \frac{(1 - x/\ell)}{2(Dt/\ell^2)^{\frac{1}{2}}} \qquad (14)$$

Stresses due to swelling were evaluated for the case of an adhesive

(Epoxide 3501) which swells by 3% in water and has a diffusion coefficient for water $D = 1·29 \times 10^{-12}$ m^{2}s^{-1}, between rigid adherends with 25 mm overlap. Normal stresses towards the edges of a joint after a series of fairly short exposures are shown in Fig. 16. Each curve shows a compressive stress at the edge, and maximum tensile stress which progresses into the joint with time. However, after an initial rise, the normal stress concentration decreases with time and so seems unlikely to contribute to long-term weakening of joints.

Sargent and Ashbee[64,65] have prepared sandwiches consisting of a thick adherend bonded to a thin sheet of glass and measured normal swelling strains induced by water by the analysis of Moiré fringes.

Kinloch *et al.* have proposed that an outer water-containing zone of an adhesive layer might constitute a crack. This suggestion has its roots in the observation that while exposure of adhesive joints to high humidity usually causes weakening, it has been frequently observed that they can withstand exposure to lower humidities for long periods without any weakening. DeLollis[66] has referred to some epoxide–aluminium joints which showed no loss of strength after exposure to laboratory humidity for up to 11 years. Kinloch *et al.*[67] exposed butt

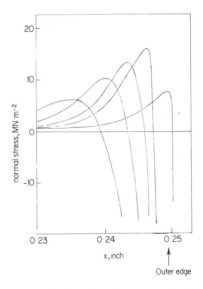

Fig. 16. *Normal stresses at the interface between adhesive and adherend for a 0·03 in glueline for moisture diffusion after 50, 1000, 2000, 5000 and 125 000 s. After Weitsman.*[63]

joints with an epoxide adhesive to 55% R.H. at 20°C for 2500 h and found no weakening; similarly Comyn et al.[2-5] found no significant weakening effect after 10 000 h at about 20°C and 45% R.H. for joints bonded with carried and uncarried variants of the adhesives FM 1000 and BSL 312. From a knowledge of water sorption isotherms of other adhesives it would be expected that joints would absorb significant quantities of water under these conditions.

Kinloch et al. proposed that there must be a critical water concentration in the adhesive layer, below which environmental weakening does not occur. The outer zone, where this critical concentration is exceeded, is regarded as a crack of length a in the bondline. Kinloch et al. tested this hypothesis with some butt joints bonded with an adhesive based on DGEBA and the 2-ethylhexanoate of 2,4,6-tris(dimethylamino)phenol, exposed to water at 20, 40, 60 and 90°C and also to air at 20°C and 55% R.H. All the water-immersed joints fell in strength and they were able to show, using a fracture mechanics approach, that strengths of the various joints could be interrelated by using a critical water concentration of 1·35%. This would mean that the outer zone which constitutes a crack has a water concentration exceeding this value, and that joints exposed to humidities too low to induce this level of water would not be weakened.

3.2. Water-induced Chemical and Physical Ageing

The failure of a water uptake plot to reach equilibrium can be taken as a sign of chemical reaction between water and adhesive. However, most systems studied in this way do in fact attain equilibrium, at ambient and moderate temperatures, so that epoxides would seem to have good water resistance. There have been some recent studies of the hydrolysis of epoxides using Fourier transform infrared spectroscopy, FTIR (see Chapter 2). Koenig et al.[68,69] examined the DGEBA–MNA and dimethylbenzylamine resin in a study which included exposure to water at 80°C for up to 90 days. Short-term effects were the hydrolysis and leaching out of unreacted anhydride but hydrolysis of ester groups occurred in samples under stress; unstressed samples were unaffected. Ester levels are shown for various exposure times in Fig. 17.

Levy, Fanter and Summers[70] subjected TGDDM–DDS to various combinations of stress up to $9\ MN\ m^{-2}$, elevated temperature and moisture in an autoclave at 135°C. Changes were monitored using

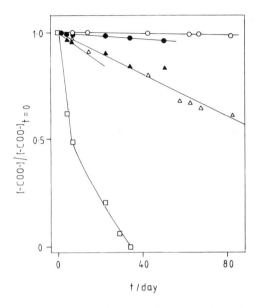

Fig. 17. Relative ester content (1744 cm^{-1} intensity) against time for an anhydride-cured epoxide resin under varying tensile loads in water (pH = 11·9) at 80°C. Stress levels: ○, zero; ●, 19; △, 26 and □, 54 MPa.[69]

FTIR. The extent of bond rupture was independent of the level of mechanical stress but was claimed to depend on swelling stresses.

The physical ageing of adhesives in the form of cracking or crazing is probably more important than hydrolysis and it seems that exposure to varying climates is an important cause of these. This has been demonstrated on many occasions by the exposure of epoxide–fibre composites to so-called thermal spikes, which are intended to simulate flight conditions of military aircraft. A supersonic dash will cause an aircraft skin temperature to rise from below zero to about 150°C as it accelerates, but as it resumes its cruising speed the temperature will have returned to below zero within a few minutes. Adhesives within the aircraft will contain some water from contact with the atmosphere, which will tend to produce voids at temperatures above boiling point.

Browning[23,60] exposed composites with a resin formed from DDS and boron trifluoride monoethylamine to thermal spikes and found equilibrium water uptake to increase with spiking. In one experiment, in which samples were subjected to one or four spikes per day, weight gains were proportional to the total number of spikes. This behaviour was attributed to micro-cracking, with water entering the cracks. The micro-cracks were demonstrated by scanning electron microscopy. Glass transition temperature was not changed by spiking, so indicating that the extra water absorbed was not contributing to plasticisation.

Mazor, Broutman and Eckstein[71] demonstrated damage to epoxide-based composites after exposure to salt or distilled water for 11 years. Samples were dried out and than re-exposed to water. Higher weight gains and diffusion coefficients shown by the 11-year samples indicated the possible occurrence of microdamage. This was supported by observed changes in mechanical properties.

Small amounts of electrolyte may be responsible for the formation of cracks or crazes, and this could be sodium chloride remaining from the reaction of epichlorohydrin with the sodium salt of bisphenol A in the production of DGEBA. The attempt by water to dilute this electrolyte could set up osmotic pressures capable of causing internal fracture. Ashbee and his co-workers[72,73] showed that the incorporation of finely ground potassium chloride into an epoxide resin led to the formation of disc-shaped cracks upon exposure to hot water. Cooling the wet epoxides to $-95 \cdot 1°C$ caused further propagation of the cracks as the water which they contained turned into ice. Fedors[74] demonstrated cracking in epoxides due to embedded sucrose or some electrolytes.

Apicella et al.[25,75,76] demonstrated Fickian uptake with an equilibrium plateau for water uptake by an adhesive based on DGEBA and TETA at 23, 45 and 75°C. However, on changing the temperature of an equilibrated sample, weight always increased and so did not move necessarily towards the level of equilibrium uptake first associated with the new temperature. For example, M_∞ for samples first equilibrated at 23, 45 and 75°C were respectively 3·92, 3·90 and 4·12%, so that if a sample was equilibrated at 75°C and the temperature then lowered to 23°C, weight loss would be anticipated. In fact, an increase was observed and this was attributed to the formation of micro-cavities. On the basis of no micro-cavities being formed at the lowest temperature, amounts of water contained in micro-cavities at 45 and 75°C were 0·23% and 0·94% respectively. Again the water apparently isolated within micro-cavities did not cause any lowering of the glass transition temperature.

Comyn et al.[26] exposed films of the epoxide DGEBA–DAPEE to five successive sorption–desorption cycles from saturated water vapour at 48°C. Their results indicated that diffusion coefficient decreased with cycling, and solubility increased slightly. For both parameters the greatest change took place during the first cycle. Again these results were ascribed as being due to micro-cracks, but a search by electron microscopy failed to reveal any.

Cracks may of course increase the rate of water transport and so lead to an apparent increase in D; they may also grow, causing an adhesive joint to fail. Water might have an effect on crack growth and so modify the strength of a joint. Mizutani and Iwatsu[77] have recently examined the effect of a number of organic liquids and water on the fracture toughness of the epoxide adhesive DGEBA–TETA using double cantilever beam specimens in short-term tests. Fracture toughness at initiation was increased with all the solvents studied and for the case of water the increase was to $0.80 \, MN \, m^{-\frac{3}{2}}$ from the value of $0.53 \, MN \, m^{-\frac{3}{2}}$ in air. Fracture toughness at initiation and arrest are shown against solvent solubility parameter in Fig. 18. Interpretation of this information depends upon the liquids being able to blunt a static crack, but when growth occurs the liquids are unable to keep up with the fast-moving crack so that at arrest, fracture toughness reverts to its value in air. The lack of correlation of fracture toughness with solvent viscosity supported this view.

Ripling, Mostovoy and Bersch[78] noted that water increased the

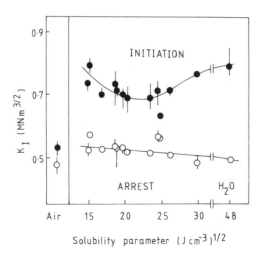

Fig. 18. The dependence of fracture toughness of DGEBA–TETA at initiation and arrest on solubility parameter of the surrounding medium.[77]

toughness of some epoxide adhesives for short loading periods, but that the load-carrying capacity at long times was decreased by water.

4. EFFECT OF WATER ON THE INTERFACE

4.1. Adsorption Theory of Adhesion and Displacement of the Adhesive by Water

Using the following argument, Kinloch and his co-workers[79,80] have demonstrated that water may displace an adhesive from the interface.

The thermodynamic work of adhesion, i.e. the work required to separate unit area of two phases in contact, is related to the surface free energies by the Dupré equation. If the phases are separated in an inert medium such as dry air the equation has the form:

$$W_A = \gamma_x + \gamma_y - \gamma_{xy} \tag{15}$$

but if separation takes place in a liquid then the form is

$$W_{AL} = \gamma_{xl} + \gamma_{yl} - \gamma_{xy} \tag{16}$$

These relationships are illustrated in Fig. 19.

Now, as discussed in Chapter 1, the surface free energy of a substance is the sum of polar and dispersion components such that

$$\gamma = \gamma^P + \gamma^D \tag{17}$$

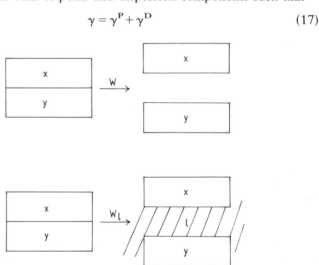

Fig. 19. *The separation of an adhesive bond in an inert medium and in a liquid.*

permitting eqns (15) and (16) to be rewritten in terms of γ^P and γ^D for each component phase. The rewritten equations are

$$W_A = 2[(\gamma_x^D \gamma_y^D)^{\frac{1}{2}} + (\gamma_x^P \gamma_y^P)^{\frac{1}{2}}] \qquad (18)$$

and

$$W_{AL} = 2[\gamma_l - (\gamma_x^D \gamma_l^D)^{\frac{1}{2}} - (\gamma_x^P \gamma_l^P)^{\frac{1}{2}} - (\gamma_y^D \gamma_l^D)^{\frac{1}{2}}$$
$$- (\gamma_y^P \gamma_l^P)^{\frac{1}{2}} + (\gamma_x^D \gamma_y^D)^{\frac{1}{2}} + (\gamma_x^P \gamma_y^P)^{\frac{1}{2}}] \qquad (19)$$

The terms γ^D and γ^P can be determined experimentally and some values of relevance to interfaces in structural adhesive joints are collected in Table 6. In Table 7 there appear some calculated values of the work of adhesion. A negative value of W_{AL} indicates that an interface is unstable in the presence of water and that water is capable of displacing adhesive from the substrate. However, the kinetics may be such that displacement by water does not occur spontaneously or even after a long period. Table 7 shows that this theory predicts that bonds between epoxides and glass, aluminium and iron or steel are thermodynamically unstable in the presence of water. However, bonds to carbon-fibre composites are stable, although a case is shown where surface treatment (CFRP treated with nitric acid) seems actually to

TABLE 6
SURFACE FREE ENERGIES (mJ m^{-2}) OF ADHERENDS, ADHESIVES AND WATER

	γ^D	γ^P	γ	Refs.
SiO_2	78	209	287	81, 82
Al_2O_3	100	538	638	83
Fe_2O_3	107	1250	1357	79
CFRP (heavily sanded)	$27\cdot4 \pm 5\cdot3$	$30\cdot6 \pm 5\cdot3$	58	84
CFRP (HNO_3-treated)	$12\cdot9 \pm 3\cdot5$	$60\cdot6 \pm 9\cdot3$	$73\cdot5$	84
Carbon fibre	$25\cdot9 \pm 1\cdot5$	$25\cdot7 \pm 3\cdot3$	$51\cdot6$	85
Epoxide[a]	$41\cdot2 \pm 2\cdot4$	$5\cdot0 \pm 0\cdot8$	$46\cdot2$	79
Modified epoxide[b]	$37\cdot2 \pm 3\cdot1$	$8\cdot3 \pm 2\cdot0$	$45\cdot5$	85
Cycloaliphatic epoxide[b]	$28\cdot9 \pm 3\cdot2$	$25\cdot0 \pm 4\cdot7$	$53\cdot9$	85
Poly (methyl methacrylate)[c]	$35\cdot9$	$4\cdot3$	$40\cdot2$	86
Water	$22\cdot0$	$50\cdot2$	$72\cdot2$	87

[a] DGEBA cured with $9\cdot4$ wt% of 2,4,6-tris(dimethylamino)phenol 2-ethylhexanoate.
[b] Cured under dry nitrogen.
[c] PMMA representing a reaction setting acrylic adhesive.

TABLE 7
WORK OF ADHESION IN AIR AND IN WATER

Interface	$W_A(mJ\,m^{-2})$	$W_{AL}(mJ\,m^{-2})$	Ref.
Epoxide/SiO$_2$	178	−57	88
Epoxide/Al$_2$O$_3$	232	−137	88
Epoxide/Fe$_2$O$_3$	291	−255	88
Epoxide/CFRP	88–90	22–44	88
Epoxide/CFRP (heavily sanded)	92	18	[a]
Epoxide/CFRP (HNO$_3$-treated)	95	−20	[a]
PMMA/Al$_2$O$_3$	216	−102	[a]
PMMA/Fe$_2$O$_3$	270	−167	[a]

[a] Calculated from data in Table 6.

reduce stability. The close agreement of the values in Table 6 for the heavily sanded CFRP composite and carbon fibres is due to the exposure of carbon fibres by the sanding process. However, although bonds to CFRP may be thermodynamically stable and indeed interfacial failure is not widely observed, loss of strength may be observed due to other processes such as plasticisation. Finally, joints using methyl methacrylate as a reaction setting adhesive are also predicted to be unstable in the presence of water.

One disadvantage with this approach is that it implies that joint strength will fall to zero, whereas in many cases strength is seen to level out at a moderate fraction of the initial dry strength. Another is that it does not allow for the recovery of strength on drying, as once it is displaced from the substrate it seems unlikely that there would be sufficient molecular mobility for the adhesive to re-establish contact with the adherend. Joints are sometimes observed to increase in strength on drying out, and the fact that complete recovery is not always reported is due in part to the fact that drying times are usually too short. The duration of exposure and drying-out times are controlled by the rate of water diffusion; for lap joints measuring about 10 mm × 10 mm bonded with epoxide adhesives, about one year is needed to reach a high level of water uptake, and a similar period for drying out. Nevertheless, the level of strength maintained during such a cycle could provide valuable information on the role of irreversible processes both within the adhesive and at the interface. Sorption–desorption cycles on joints that have appeared in the literature have mostly been of fairly short duration and have generally shown that some recovery can take place, although it seems doubtful whether all water was removed in the drying stage. Comyn *et al.*[2,5] exposed 25 mm × 12·5 mm double lap joints with FM 1000 and BSL 312 adhesives to 1000 h exposure followed by 1000 h drying. Some strength recovery occurred in all cases. Ishii and Yamaguchi[89] exposed butt joints to water at temperatures between 20 and 100°C for up to 100 h. Joints bonded with an epoxide–polyamide adhesive fell in strength on exposure between 60 and 100°C and some bonded with a nitrile phenolic adhesive fell in strength on exposure at 100°C. These strengths were only slightly recovered on drying out at 40°C.

A further problem is that the surface of an adhesive which has been cured against a substrate may be different from one which has been cast against air, and hence the values of γ^D and γ^P on the air-cast samples may not reflect the behaviour of an adhesive joint. Differences may arise from orientation or preferential adsorption of one adhesive component at the interface, chemical bond formation across the interface or mechanical interlocking[90] of the substrate and adhesive. A recent examination[91] of the aluminium/epoxide interface by inelastic electron tunnelling spectroscopy (see Chapter 2) gave evidence for the possibility of the chemical adsorption of amine hardeners on aluminium oxide.

4.2. Primary Forces across the Interface

In the absence of much experimental evidence, it is easy to speculate that forces other than secondary force attractions operate across adhesive interfaces. As shown in Chapter 1 (Table 3), primary forces include covalent bonds and ionic attractions between ion-pairs.

Some encouraging evidence for covalent bonding was provided by a study of mild steel surfaces primed with silane coupling agents.[44] Using secondary ion mass spectrometry the presence of the $FeSiO^+$ radical was detected for some systems, and the environmental resistance of adhesive joints could be directly related to the occurrence of this species (see Chapter 1, Figs 7 and 8). The presence of covalent bonds across the interface may be a factor in the superior durability of phenolic-based adhesives. Such bonds might be formed by the coordination of phenolic oxygens to aluminium ions, as has been observed by the formation of complexes between aluminium and 8-hydroxyquinoline[92] or alizarin dyes.[93,94] Further support for this view comes from inelastic electron tunnelling spectra of phenol and hydroquinone supported on a thin layer of aluminium oxide, which were found to adsorb predominantly as $C_6H_5O^-$ or $HOC_6H_4O^-$.[95]

Ion-pairs at the interface can be used to account for the effect of water on adhesive joints. Ions might arise at the interface in structural adhesive joints because of the metal preparation procedure and the mechanism of epoxide cure; metals are subjected to surface treatments which give them a surface layer of an ionic oxide and the mechanism of epoxide ring opening is ionic. An amine group will add to an epoxide group in the following manner:

$$\underset{R}{\overset{R}{\diagdown}}N: \; + \; CH_2\text{---}CH\text{---} \;\longrightarrow\; \text{---}\overset{+}{N}R_2\text{---}CH_2\text{---}\overset{O^-}{\underset{|}{C}H}\text{---}$$

Where the amine is tertiary ($R \neq H$) anionic polymerisation will proceed on the alkoxide ion. In the case of primary or secondary amines where one or both of the R groups is a hydrogen atom and polymerisation is by condensation, proton transfer will occur to eliminate the zwitterion thus:

$$\text{---}\overset{+}{N}H_2\text{---}CH_2\text{---}\overset{O^-}{\underset{|}{C}H}\text{---} \;\longrightarrow\; \text{---}NH\text{---}CH_2\text{---}\overset{OH}{\underset{|}{C}H}\text{---}$$

These are the processes which will normally occur in the bulk of the adhesive. At the interface, however, ion exchange can occur thus:

$$\text{---}\overset{+}{N}H_2\text{---}CH_2\text{---}\overset{O^-}{\underset{|}{C}H}\text{---} \; + \; \diagup\!\!\!\overset{\diagdown}{Al^+X^-} \;\longrightarrow\; \diagup\!\!\!\overset{\diagdown}{Al^+O^-}\text{---}\underset{|}{C}H\text{---}CH_2\text{---}\underset{|}{\overset{+}{N}}H_2X^-$$

so providing an electrostatic attractive force across the interface.

Coulombic attraction has been suggested as contributory in binding acid dyes to anodised aluminium.[96-98] Acid dyes are more commonly used for dyeing wool and nylon and they contain one or more $-SO_3^-$ groups. Coulombic attraction occurs between these and cationic groups ($-\overset{+}{N}H_3$, $-CO\overset{+}{N}H_2{}^-$) in the fibre. The production of ion exchange sites on alumina has been ascribed to the acid treatment of aluminium surfaces,[99] e.g.

$$\diagup\!\!\!\overset{\diagdown}{Al}\text{---}OH \; + \; HCl \;\longrightarrow\; \diagup\!\!\!\overset{\diagdown}{Al^+Cl^-} \; + \; H_2O$$

Surface treatments for aluminium involving acid solutions, such as etching or anodising in chromic–sulphuric or phosphoric acid, are known to give strong bonds with the best durability and the reason for this could be due to some extent to the fact that they produce exchangeable ions on alumina.

The force F between two ions is given by the expression

$$F = \frac{q_1 q_2}{4\pi\kappa\varepsilon_0 r^2} \tag{20}$$

Pauling[100] gives the crystal radii of Al^{3+} and O^{2-} ions as $0\cdot050$ and $0\cdot140$ nm, so that if there is just one positive charge on aluminium available for attracting the alkoxide, the other two being neutralised by oxide ions in alumina, and the relative permittivity of the medium is 4, then the force between the ions at the interface will be $1\cdot60\times10^{-9}$ N. Should there be one ion-pair every 1 nm on a square lattice surface, then the force of attraction would be 1600 MN m^{-2}, which is much greater than observed joint strengths. Of course the ion-pairs could be less plentiful than this, but the calculation does illustrate the feasibility of ionic attraction holding joints together.

What is the role which water will play in this electrostatic model? It will enter the adhesive and increase the relative permittivity at the interface, and so weaken the electrostatic force. Quite simply the interionic forces with wet and dry interfaces will be related by

$$\frac{F_{\text{wet}}}{F_{\text{dry}}} = \frac{\kappa_{\text{dry}}}{\kappa_{\text{wet}}} \tag{21}$$

Delmonte[101] reported the relative permittivity of epoxides cured with DETA, MPD and phthalic anhydride as $4\cdot6$ and values of about 4 have been reported for TETA cured epoxide.[102] Kadotani[43] reported a value of $2\cdot96$ for an anhydride-cured cycloaliphatic epoxide. All these values are at room temperature, where the relative permittivity of water is about 80.[103] Thus the addition of a small amount of water could cause the relative permittivity of an epoxide to increase significantly. Information on the relative permittivities of mixtures of organic solvents with water shows that κ varies approximately linearly with the composition of the mixture, examples demonstrating this being dioxan–water[104] and methyl cellosolve–water[105] mixtures. It seems reasonable to assume that the same might be true for epoxide–water mixtures if water molecules within epoxides are free to orientate themselves in ionic electrical fields. Hence, if an epoxide absorbs between 2 and 5% of water,[14] its permittivity might be increased from about 4 to between $5\cdot5$ and 8, and the force between the ions at the interface reduced to between 70 and 50% of the dry value. Although it may be fortuitous this reduction is in good agreement with values for the data presented in Figs 12, 13 and 14, where the strength retentions are 80% and 54% for Fig. 12, 59% and 47% for Fig. 13, and 51% if the data in the linear region of Fig. 14 are extrapolated to saturation. A strength retention of 65% has been observed[5] for some joints using the modified epoxide adhesive BSL 312. Thus the advantages of considering ion-pairs at the interface as major contributors to the strength of epoxide to aluminium bonds are that (i) a reduction in strength is predicted which is similar to observed strength losses, (ii) partial weakening rather than total strength loss is indicated and (iii) strength should be recoverable on removal of water.

Could this idea be extended to other adhesives and substrates? Other metallic adherends present no problems as they, like aluminium, are covered with an oxide layer. Other structural adhesives may be based on phenolic or acrylic compounds where the phenolic —OH group could form a salt with a substrate cation; Fowkes and Mostafa have proposed acid–base interactions for the adsorption of poly(methyl methacrylate) on silica.[106]

Water is the most important fluid in joint durability, but other liquids may throw light on the validity of opposing theories. Kerr, MacDonald and Orman[107] found that whilst ethanol severely deteriorated the mechanical properties of an epoxide adhesive, its effect on joint strengths was slight. Water, on the other hand, severely weakened joints but had little effect on the adhesive. This behaviour can be ascribed to water but not ethanol having the ability to displace the adhesive, but the ion-pairing approach also offers an explanation because of the low relative permittivity ($\kappa = 24 \cdot 2$)[108] of ethanol. The thermodynamic displacement theory also predicts[79] that formamide should displace an epoxide adhesive from a metal oxide, but the high relative permittivity of this fluid ($\kappa = 109$)[108] also would imply a large weakening effect by the ion-pair theory.

One problem with the ion-pair theory is in calculating the change in relative permittivity induced by water. This is because the ion-pairs are at an interface rather than being immersed in an isotropic medium. Also, water may not be evenly distributed throughout the adhesive; it may be preferentially adsorbed at the interface and so cause even greater weakening or it may be isolated in clusters[26] within the adhesive and so be removed from the ion-pairs. Alternatively, the ion-pairs may themselves induce cluster formation,[109] in which case weakening would be increased. There also exists the possibility that relative permittivity is reduced (relative to the bulk) in the close vicinity of ion-pairs,[110] but this would influence both the wet and dry adhesive in a similar manner so that the fractional weakening on water uptake would not be affected. However, it has also been indicated that the relative permittivity of water can be much greater than 80 in thin films[111] or in close association with polymers.[112]

5. EFFECT OF WATER ON THE ADHEREND

Reference has already been made to water attacking adherends consisting of fibres in plastic matrices, but with metallic adherends water may (i) attack the oxide layer or (ii) be involved in corrosion of the bulk adherend. In most cases gross corrosion of a metallic substrate is not a problem and its occurrence in failed joints is often taken as a sign of post-debonding corrosion[79] rather than as a primary cause of failure. Possible exceptions to this are when the environment is particularly corrosive, e.g. salt water, or when two different metals are in electrical contact within an adhesive, as is the case with aluminium alloy clad with pure aluminium.[113] Here differences in electrode potential will give the possibility of one metal being preferentially corroded.

The choice of surface treatment given to a metal before adhesive bonding has an important affect on the performance of a joint in wet conditions, discussed in detail in the following chapters. For example, with aluminium and its alloys etching or anodising in chromic or phosphoric acid usually gives the best performance. It seems that differences in the chemistry and morphology of the oxides generated by the different treatments affect their stability in the presence of water.

Noland[114] has shown that the structure of aluminium oxide may change on exposure to water. This was based on an observed change in the binding energy of the aluminium $2p$ peak when an oxide produced by etching in chromic acid solution was exposed to water vapour. In contrast the oxide produced by anodising in phosphoric acid was stable to water with hardly any change in the aluminium $2p$ peak. These results, of course, are in keeping with the superior durability attainable with phosphoric-anodised aluminium adherends (see Chapter 5).

It is not clear why different pre-treatments give oxides of varying stabilities but some interesting proposals are beginning to emerge which are based on the role of impurities in the oxide. Specifically, phosphate ions remaining from treatments in phosphoric acid may inhibit hydration[115,116] (see Chapters 2 and 5) whilst the presence of copper[117] or magnesium[116] impurities may decrease oxide stability. Indeed, Kinloch[118] used XPS to measure magnesium levels and found a marked correlation between this parameter and joint durability.

6. EFFECT OF STRESS

Adhesive joints generally have to bear loads at least sometime during their working life and this of course may have an effect both on the kinetic and mechanistic processes which have been discussed above. Firstly, stress might modify the rate of water diffusion. However, the evidence which exists indicates that this is not likely to be a large effect. Kim and Broutman[119] measured diffusion coefficients in some stressed tensile bars of graphite–epoxide composites. The level of stress was 25% of ultimate and exposure was to liquid water. Some of their results are collected in Table 8, and it can be seen that D increases with stress, whilst there is a small increase in M_∞. Interestingly, the activation energy for diffusion was not significantly altered by the application of stress, but the pre-exponential factor in the Arrhenius equation, D_0, was increased. The enhanced diffusion rate for stressed samples could be due to stress causing microscopic defects of interfacial separation.

Gillat and Broutman[120] undertook a similar experiment but went to

TABLE 8

DIFFUSION COEFFICIENTS AND SOLUBILITIES OF WATER IN SOME STRESSED AND UNSTRESSED GRAPHITE–EPOXIDE COMPOSITES. AFTER KIM AND BROUTMAN[119]

$Temp(°C)$	$D/10^{-14}\ m^2 s^{-1}$		$M_\infty(\%)$	
	unstressed	stressed	unstressed	stressed
25	4·2	7·8	1·70	1·76
40	13·3	24·4	1·66	1·82
60	40·8	50·7	1·69	1·80
80	45·7	—	1·71	—
E_d (kJ mol^{-1})	49·0	49·5		
$D_0/10^{-5}\ m^2 s^{-1}$	1·68	3·7		

TABLE 9
DEPENDENCE OF DIFFUSION COEFFICIENT AND SOLUBILITY
OF WATER IN STRESSED GRAPHITE-EPOXIDE COMPOSITES
AT 333 K. AFTER GILLAT AND BROUTMAN[120]

Stress		$D/10^{-13} m^2 s^{-1}$	$M_\infty(\%)$
(% of ultimate)	*(MPa)*		
0	0	1·95	1·75
25	172	3·86	1·60
45	306	3·70	1·65
65	436	11·67	1·75

65% of ultimate stress of some epoxide–graphite composites. Uptake remained Fickian at all levels of stress, but whilst equilibrium uptake was not significantly altered by stress, there was a tendency for D to increase (Table 9). In an analysis based upon the free volume theory, Fahmy and Hurt[121] have shown that D should increase exponentially with stress.

The effect of preloading has been examined by Morgan, O'Neal and Fanter,[122] who exposed tensile bars of TGDDM–DDS to a constant load between 0 and 60 MPa for 1 h at room temperature and then examined their moisture uptake. The values of D showed no trend with applied stress but the equilibrium uptake levels rose from *ca* 4·5% at prestress levels between 0 and 40 MPa to *ca* 5·0% at higher levels.

The effect of stress upon double lap joints exposed to saturated air at 50°C has been examined[2,5] for some adhesives based on FM 1000 and BSL 312. Joints were stressed to 20% of their mean initial dry strength and exposed for periods up to 10 000 h. Joints became weaker during exposure, but stressed and unstressed joints were weakened to the same extent, which points to the diffusion coefficient of water having the same value in both stressed and unstressed joints. A few stressed joints did fail during exposure, but other members of their set which remained intact had strengths comparable with unstressed joints.

Cotter[1] has shown a number of strength versus time plots for aluminium double lap joints with six adhesives described as epoxide–Novolak, modified epoxide, nitrile–phenolic, epoxide–polyamide, vinyl-phenolic and epoxide–phenolic exposed for up to six years to either a hot/wet, hot/dry or temperate natural climate. Some joints were exposed without stress but others were stressed to 20% of their mean initial dry strength (15% in the case of the vinyl–phenolic joints in the hot climates). Most joints did not show much sensitivity to stress, the exceptions being the epoxide–polyamide joints in all climates, the modified epoxide in the temperate climate and the vinyl–phenolics in hot/wet surroundings. A factor which may lead to stress insensitivity is the possible occurrence of a critical stress level below which no environmental attack takes place.[123]

Structural adhesives depend upon covalent bonds for their cohesive strength and also the possibility exists of covalent bonds across the interface. Any chemical reaction involving the destruction of these bonds would be accelerated if the bonds were stressed. The rate of destruction would increase with stress in accordance with eqn (22).

$$k = A \exp \left(\frac{-(E_a - v^* \sigma)}{\mathbf{R}T} \right) \tag{22}$$

Effectively the bonds are activated by stress with an apparent lowering of the Arrhenius activation energy, E_a. The equation has been verified for the mechanical destruction of polypropylene and linear and branched polyethylenes by Zhurkov and Korsukov,[124] who give values of 2·1, 1·1 and 2·4 kJ mol^{-1} MPa^{-1} for the respective values of the constant v^*. Further, it has been used by Antoon and Koenig[69] to fit the rates of hydrolysis of an anhydride-cured epoxide resin under stress (the data appear as our Fig. 17) and here the value of v^* is *ca* 1·2 kJ mol^{-1} MPa^{-1}.

7. CONCLUDING REMARKS

Most detailed studies on the kinetics of adhesive joint durability have employed adherends which have been carefully prepared, and in many cases given a surface treatment known to confer good durability. In these cases the kinetics of strength loss are controlled by the rate of diffusion of water in any permeable phase; this will always be the adhesive itself but should one or both adherends be a plastic composite these will also be permeable. Water which has entered by diffusion can modify the properties of the adhesive by both reversible (plasticisation, swelling) and irreversible (cracking and crazing) processes.

Van der Waals forces will always be present at an adhesive/substrate interface and their displacement by water is clearly predicted for most substrates. These forces are sufficient to account for observed joint strengths, but invocation would imply virtually complete loss of joint strength without recovery in the presence of water. The advent of newer surface analytical techniques such as inelastic electron tunnelling spectroscopy and electron spectroscopy for chemical analysis may in the future provide evidence for the presence of intrinsically stronger and more durable ion-pair and covalent interactions across interfaces.

REFERENCES

1. Cotter, J. L., in *Developments in Adhesives—1*, ed. W. C. Wake, Applied Science Publishers, London (1977), chapter 1.
2. Brewis, D. M., Comyn, J., Cope, B. C. and Moloney, A. C., *Polymer*, **21** (1980), 344.
3. Brewis, D. M., Comyn, J. and Tegg, J. L., *Int. J. Adhes. Adhes.* **1** (1980), 35.
4. Brewis, D. M., Comyn, J., Cope, B. C. and Moloney, A. C., *Polymer*, **21** (1980), 1477.
5. Brewis, D. M., Comyn, J., Cope, B. C. and Moloney, A. C., *Polym. Eng. Sci.*, **21** (1981), 797.
6. Baker, F. S., *J. Adhes.*, **10** (1979), 107.
7. Butt, R. I. and Cotter, J. L., *J. Adhes.*, **8** (1976), 11.
8. Zalucha, D. J., *Natl SAMPE Tech. Conf.*, (1981), 92.
9. Antoon, M. K. and Koenig, J. L., *J. Macromol. Sci.—Rev. Macromol. Chem.*, **C19**, (1980), 135.
10. Shirrell, C. D. and Halpin, J., *ASTM Spec. Tech. Publ.*, **STP617** (1977), 514.
11. Crank, J., *The Mathematics of Diffusion*, 2nd edn., Oxford University Press (1975).
12. Carslaw, H. S. and Jaeger, J. C., *Conduction of Heat in Solids*, 2nd edn, Oxford University Press (1959).
13. Shen, C-H. and Springer, G. S., *J. Compos. Mater.*, **10** (1976), 2.
14. Brewis, D. M., Comyn, J., Shalash, R. J. A. and Tegg, J. L., *Polymer*, **21** (1980), 357.
15. Daniely, N. D. and Long, E. R., Jr, *J. Polym. Sci., Chem. Ed.*, **19** (1981), 2443.
16. Marom, G. and Broutman, L. J., *J. Appl. Polym. Sci.*, **26** (1981), 1493.
17. Fujita, H., *Adv. Polym. Sci.*, **3** (1961), 1.
18. Kaelble, D. H. and Dynes, P. J., *Mater. Eval.*, **35** (1977), 103.
19. McKague, E. L., Reynolds, J. D. and Halkias, J. E., *J. Appl. Polym. Sci.*, **22** (1978), 1643.
20. McKague, E. L., Reynolds, J. D. and Halkias, J. E., *Trans. ASME.*, **98** (1976), 92.
21. Cook, T. S., Walrath, D. E. and Francis, P. H., *Natl SAMPE Symp. Exhib.*, **22** (Diversity-Technol. Explos.) (1977), 339.
22. Long, E. R., *NASA Technical Paper* 1474 (1979).

23. Browning, C. E., *Polym. Eng. Sci.*, **18** (1978), 16.
24. Illinger, J. L. and Schneider, N. S., *Polym. Eng. Sci.*, **20** (1980), 310.
25. Apicella, A., Nicolais, L., Astarita, G. and Drioli, E., *Polymer*, **20** (1979), 1143.
26. Brewis, D. M., Comyn, J. and Tegg, J. L., *Polymer*, **21** (1980), 134.
27. Manson, J. A. and Chiu, E. H., *J. Polym. Sci., Symposium* No. 41 (1973), 95.
28. Meares, P., *J. Amer. Chem. Soc.*, **76** (1954), 3415.
29. Sandorff, P. E. and Tajima, Y. A., *SAMPE Q.*, **10** (1979), 21.
30. Musty, J. W. G., Pattle, R. E. and Smith, P. J. A., *J. Appl. Chem.*, **16** (1966), 221.
31. Augl, J. M. and Berger, A. E., *Natl SAMPE Tech. Conf.*, **8** (1976), 383.
32. Perera, D. Y. and Heertjes, P. M., *J.O.C.C.A.*, **54** (1971), 395.
33. Mozisek, M., *Jad. Energ.*, **22** (1976), 448.
34. Moy, P. and Karasz, F. E., *Polym. Eng. Sci.*, **20** (1980), 315.
35. Althof, W., *Proc. 11th Natl SAMPE Tech. Conf.* (1979), 309.
36. Althof, W., *DFVLR-Forschungsbericht* 79-06 (1979).
37. Althof, W., *Aluminium*, **55** (1979), 600.
38. Brewis, D. M., Comyn, J., Moloney, A. C. and Tegg, J. L., *Europ. Polym. J.*, **17** (1981), 127.
39. Schulte, R. L. and DeIasi, R. J., *IEEE Trans. Nucl. Sci.*, **NS28** (1981), 1841.
40. Parker, B. M., *Proc. Conf. Jointing in Fibre Reinf. Plastics, London* (1978), 95.
41. Bowditch, M. R. and Stannard, K. J., *Adhesion—5*, ed. K. W. Allen, Applied Science Publishers, London (1981), p. 93.
42. Laird, J. A., *Final report, Navy contract W-0679-C(FBM)* (1963).
43. Kadotani, K., *Composites*, **11** (1980), 199.
44. Gettings, M. and Kinloch, A. J., *J. Mater. Sci.*, **12** (1977), 2511.
45. Gettings, M., Baker, F. S. and Kinloch, A. J., *J. Appl. Polym. Sci.*, **21** (1977), 2375.
46. Boerio, F. J., Cheng, S. Y., Armogan, L., Williams, J. W. and Gosselin, C., *Proc. 35th Ann. Conf. Reinf. Plast./Compos.*, Soc. Plast. Ind. (1980), 23C.
47. Boerio, F. J. and Williams, J. W., *Applicns Surf. Sci.*, **7** (1981), 19.
48. Bascom, W. D., *Macromolecules* **5** (1972), 792.
49. Sung, N. H., Ni, S. and Paik Sung, C. S., *Org. Coat. Plast. Chem.* (1980), 743.
50. Sung, N. H. and Paik Sung, C. S., *Proc. 35th Ann. Conf. Reinf. Plast./Compos.*, Soc. Plast. Ind. (1980), 23B.
51. Barrie, J. A. and Machin, D., *J. Macromol. Sci.*, **B3** (1969), 645.
52. Kelley, F. N. and Beuche, F., *J. Polym. Sci.*, **50** (1961), 549.
53. Fox, T. G., *Bull. Amer. Phys. Soc.*, **1** (1956), 123.
54. McMillan, J. A. and Los, S. C., *Nature*, **206** (1965), 806.
55. Sugisake, M., Suga, H. and Seki, S., *Bull. Chem. Soc. Japan*, **41** (1968), 2591.
56. Rasmussen, D. H. and MacKenzie, A. P., *J. Phys. Chem.*, **75** (1971), 967.
57. Frank, R., *Water*, Plenum Press, New York (1972).

58. Morgan, R. J. and Mones, E. T., *11th Natl SAMPE Tech. Conf.* (1979), 218.
59. Browning, C. E., Husman, G. E. and Whitney, J. M., *ASTM Spec. Tech. Publ.* 617 (1977), 481.
60. Browning, C. E., *Natl SAMPE Symp. Exhib.*, **22** (1977), 365.
61. Shalash, R. J. A., Ph.D. Thesis, Leicester Polytechnic (1980).
62. Gazit, S., *J. Appl. Polym. Sci.*, **22** (1978), 3547.
63. Weitsman, Y., *J. Compos. Mater.*, **11** (1977), 378.
64. Sargent, J. P. and Ashbee, K. H. G., *J. Adhesion*, **11** (1980), 175.
65. Sargent, J. P. and Ashbee, K. H. G., *J. Phys. D*, **14** (1981), 1933.
66. DeLollis, N. J., *Natl SAMPE Symp. Exhib.*, **22** (1977), 673.
67. Gledhill, R. A., Kinloch, A. J. and Shaw, S. J., *J. Adhesion*, **11** (1980), 3.
68. Antoon, M. K., Koenig, J. L. and Serafini, T., *J. Polym. Sci., Physics Edn*, **19** (1981), 1567.
69. Antoon, M. K. and Koenig, J. L., *J. Polymer Sci., Physics Edn*, **19** (1981), 197.
70. Levy, R. L., Fanter, D. L. and Summers, C. J., *J. Appl. Polym. Sci.*, **24** (1979), 1643.
71. Mazor, A., Broutman, L. J. and Eckstein, B. H., *Natl Tech. Conf. SPE (Prepr.)* (1976), 77.
72. Farrar, N. R. and Ashbee, K. H. G., *J. Phys. D*, **11** (1978), 1009.
73. Nicholas, J. and Ashbee, K. H. G., *J. Phys. D*, **11** (1978), 1015.
74. Fedors, R. F., *Polymer*, **21** (1980), 713.
75. Apicella, A. and Nicolais, L., *Ind. Eng. Chem. Prod. Res. Dev.*, **20** (1981), 138.
76. Apicella, A., Nicolais, L., Astarita, G. and Drioli, E., *Polym. Eng. Sci.*, **21** (1981), 17.
77. Mizutani, K. and Iwatsu, T., *J. Appl. Polym. Sci.*, **26** (1981), 3447.
78. Ripling, E. J., Mostovoy, S. and Bersch, C., *J. Adhesion*, **3** (1971), 145.
79. Gledhill, R. A. and Kinloch, A. J., *J. Adhesion*, **6** (1974), 315.
80. Kinloch, A. J., Dukes, W. A. and Giedhill, R. A., *Polym. Sci. Technol.*, **9B** (1975), (Adhes. Sci. Technol.), 597.
81. Boyd, G. E. and Livingstone, H. K., *J. Amer. Chem. Soc.*, **64** (1942), 2383.
82. Shartsis, L. and Spinner, S., *J. Res. Natl Bur. Stand.*, **46** (1951), 385.
83. Schonhorn, H., *J. Phys. Chem.*, **71** (1967), 4578.
84. Crane, L. W., Hamermesh, C. L. and Maus, L., *SAMPE J.*, **12** (1976), 6.
85. Lewis, A. F. and Gounder, R. T. N., *Treatise on Adhesion and Adhesives*, Vol. 5, ed. R. L. Patrick, Marcel Dekker, New York (1981), p. 349.
86. Owens, D. K. and Wendt, R. C., *J. Appl. Polym. Sci.*, **13** (1969), 1741.
87. Fowkes, F. M., *Treatise on Adhesion and Adhesives*, Vol. 1, ed. R. L. Patrick, Marcel Dekker, New York (1967), p. 325.
88. Kinloch, A. J., *J. Adhesion*, **10** (1979), 193.
89. Ishii, H. and Yamaguchi, Y., *Kogakuin, Daigaku Kenkyu Hokoku*, **44** (1978), 652.
90. Chang, W. V. and Wang, J. S., *J. Appl. Polym. Sci.*, **26** (1981), 1759.
91. Comyn, J., Horley, C. C., Oxley, D. P., Pritchard, R. G. and Tegg, J. L., *J. Adhesion*, **12** (1981), 171.

92. Brockmann, W., *Adhes. Age,* June (1977), 30.
93. Allen, R. L. M., *Colour Chemistry,* Nelson, London (1971).
94. Kiel, E. G. and Heertjes, P. M., *J. Soc. Dyers Colorists,* **79** (1963), 21, 61, 186.
95. Lewis, B. F., Bowser, W. M., Hurn, J. L., Luu, T. and Weinberg, W. H., *J. Vac. Sci. Technol.,* **11** (1974), 262.
96. Giles, C. H., Mehta, H. V., Stewart, C. E. and Subramanian, R. V. R., *J. Chem. Soc.* (1954), 4360.
97. Giles, C. H., *Rev. Prog. Coloration,* **5** (1974), 49.
98. Cummings, T., Craven, H. C., Giles, C. H., Rahman, S. M. K., Sneddon, J. G. and Stewart, C. E., *J. Chem. Soc.* (1959), 535.
99. O'Connor, D. J., Johansen, P. G. and Buchanan, A. S., *Trans. Faraday Soc.,* **52** (1956), 229.
100. Pauling, L., *The Nature of the Chemical Bond,* 3rd edn. Cornell University Press (1960).
101. Delmonte, J., *J. Appl. Polym. Sci.,* **2** (1959), 108.
102. Daly, J. and Pethrick, R. A., *Polymer,* **22** (1981), 37.
103. Malmbery, C. G. and Maryott, A. A., *J. Res. Nat. Bur. Standards,* **56** (1956), 1.
104. Critchfield, F. E., Gibson, J. A. and Hall, J. C., *J. Amer. Chem. Soc.,* **75** (1953), 1991.
105. Atkinson, G. and Tsubota, H., *J. Amer. Chem. Soc.,* **88** (1966), 3901.
106. Fowkes, F. M. and Mostafa, M. A., *Ind. Eng. Chem. Prod. Res. Dev.,* **17** (1978), 3.
107. Kerr, C., MacDonald, N. C. and Orman, S., *J. Appl. Chem.,* **17** (1967), 62.
108. Robinson, R. A. and Stokes, R. H., *Electrolyte Solutions,* Butterworth, London (1955).
109. Starkweather, H. W., *Polym. Prepr., Amer. Chem. Soc. Div. Polym. Chem.,* **16** (1975), 740.
110. Barker, R. E. and Sharbaugh, A. H., *J. Polymer. Sci., Part C,* **10** (1965), 139.
111. Palmer, L. S., Cunliffe, A. and Hough, J. M., *Nature,* **170** (1952), 796.
112. Jacobsen, B., *J. Amer. Chem. Soc.,* **77** (1955), 2919.
113. Riel, F. J., *SAMPE J.,* **7** (1971), 16.
114. Noland, J. S., in *Adhesion Sci. Technol.,* ed. L. H. Lee, Plenum Press, New York (1975), p. 413.
115. Sun, T. S., McNamara, D. K., Ahearn, J. S., Chen, J. M., Ditchek, B. and Venables, J. D. *Appl. Surf. Sci.,* **5** (1980), 406.
116. Kinloch, A. J. and Smart, N. R., *J. Adhes.,* **12** (1981), 23.
117. Sun, T. S., Chen, J. M., Venables, J. D. and Hopping, R., *Appl. Surf. Sci.,* **1** (1978), 202.
118. Kinloch, A. J., *Adhesion—6,* ed. K. W. Allen, Applied Science Publishers, London (1982), p. 95.
119. Kim, R. H. and Broutman, L. J., *4th Int. Conf. Deformation, Yield and Fracture in Polymers,* Cambridge (1971).
120. Gillat, O. and Broutman, L. J., *ASTM, STP* No. 685 (1978), p. 61.
121. Fahmy, A. A. and Hurt, J. C., *Polym. Compos.,* **1** (1980), 77.

122. Morgan, R. J., O'Neal, J. E. and Fanter, D. L., *J. Mater. Sci.*, **15** (1980), 751.
123. Cherry, B. W. and Thomson, K. W., *Adhesion—1*, ed. K. W. Allen, Applied Science Publishers, London (1977), p. 251.
124. Zhurkov, S. N. and Korsukov, V. E., *Sov. Phys. Solid State*, **15** (1974), 1379.

Part II
MATERIALS

4

Adhesives

J. D. MINFORD

Aluminum Company of America, Alcoa Center, Pennsylvania, USA

1. INTRODUCTION

That adhesion science remains a controversial subject is demonstrated by the fact that no universal theory of adhesion exists.[1] Many models have been proposed to explain adhesion, of which refs 2–16 are a partial listing. According to Good[17] all attempts to define the term 'adhesion' that have been proposed in the literature would be as lengthy as any article on the subject. In addition, all definitions would probably still be unsatisfactory for one or more fundamental reasons. The ASTM definition may be generally acceptable, that adhesion is the state in which two surfaces are held together by interfacial forces which may consist of valence forces or interlocking action, or both. We might propose the additional thought that because of these interacting forces, separation is resisted. One of these bodies may be considered to be the adhesive while the other will be referred to as the adherend. The presence of an adhesive between two adherends will then constitute an adhesive joint.

The definition of a structural adhesive varies from observer to observer depending upon his view of what the adhesive in a joint should accomplish. As defined in Chapter 1, common requirements are that the joint be relatively permanent and load-bearing. A favourable balance between rigidity and overall high strength is also usually desirable.

It has been mentioned that there are both mechanical and chemical contributions to the adhesion between adhesive and adherend. Mechanical adhesion is most easily visualised, as shown in the anchoring of an organic polymer in a rough surface profile as in Fig. 1. Obviously, this so-called hooking effect operates maximally where

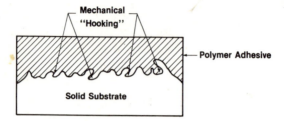

Fig. 1. Schematic representation of mechanical 'hooking'.

porous adherends are joined. Opportunity for mechanical interlocking between adhesive and adherend, however, can range from those surfaces viewed as macroscopically rough to those microscopic surface textures described by Bijlmer[18] as micro-etch pitting on metals. Evans and Pǎckham[19] have recently demonstrated significant improvements in peel strength to roughened surfaces, and Eich *et al.*[20] have shown the significant influence of roughness on wetting and adhesion in dental adhesive applications.

The chemical attributes to adhesion cover a much wider and more subtle range of interactions which can be described as primary or secondary chemical bridges. Primary bridging results from direct chemical reaction between adhesive and adherend while secondary bridging involves the residual electrical forces which surround the respective molecules. These fundamental energy relationships, their association with the wetting phenomena, and the various mechanisms of adhesion that have been proposed have already been discussed in Chapter 1. They will not be repeated here, so that more attention may be devoted to separate consideration of the various structural adhesive families and the durability data on them that exist in the literature.

The forces of mutual attraction between adhesives and adherends can be spectacular when the adhesive molecules are large, as in modern adhesive formulations. For structural adhesives, however, more is required than their ability to link with the adherend. The cohesive strength of both must also be sufficient to carry and transmit the applied loads. Thus, the adhesive cannot be studied in isolation when considering the durability of structural joints. The multiplicity of modes of possible joint failure must be identified, and the role the adhesive plays in creating a weakest link in the overall joint has to be established. It is important to recognise that a bonded joint is a construction rather than a single material. The adhesive as one part of the construction can effect joint failure under service conditions in a variety of ways. There may be destruction of the adhesive leading to a cohesive-type failure. With a structural tape adhesive the failure site may be between carrier and adhesive, which might still be considered cohesive in nature. The separation may be between the adhesive and adherend, which is generally described as an interfacial, or adhesive-type, failure. Finally, the adherend surface may be destroyed by gross corrosion or more subtle changes in the oxide (invariably present on metallic adherends) which are not associated with failure of the adhesive. It will be the primary aim of this chapter to summarise the present state of the art as regards the contribution of the adhesive to joint durability and long-term service life.

2. ADHESIVE MATERIALS AND PROPERTIES

The task of classifying structural adhesives is complicated by the fact that most successful proprietary formulations consist of several chemical types. Usually, several basic adhesive-like substances have been combined into a heterogeneous mixture having stable properties.[21] McGuire[22] first separates the raw materials for adhesives into the categories of naturally occurring and synthetically constituted materials. Each basic material will have certain advantages as an adhesive but also certain limitations. For structural purposes, the property limitations of the animal-origin materials almost preclude their use. Of maximum importance for structural adhesive formulation are a few synthetic elastomers and thermoplastic synthetic resins mixed with a variety of thermosetting synthetic resins. The most common members of these groups are neoprene and acrylonitrile elastomers; acrylic, nylon or vinyl thermoplastic resins; and epoxy, phenolic, polyurethane and silicone thermosets. References 23–32 provide further reading about the variety of structural adhesives available. For an excellent technical summary of the composition and performance characteristics of the most important types of structural adhesives, the author recommends the chapter on this subject by Bolger.[33]

2.1. Polymer/Adhesive Interrelationship

For a complete understanding of the interrelationships between polymers and adhesives, it would be necessary to study the factors affecting the ultimate strength properties of solids in general and polymers in particular. This is obviously beyond the intended scope of this chapter; there is a variety of publications on the subject to which reference may be made.[34–39]

It is pertinent, however, to consider the relationship of varying temperature on the volume properties of polymers and the glass transition temperature, T_g. The mechanical properties of uncrosslinked and unfilled polymers change dramatically around this temperature. Below T_g the glassy polymer cannot elongate significantly when stressed, while above this point elongations of several hundred per cent may be possible. Under this latter condition there is no well-defined failure stress. When polymers are crosslinked and loaded with inorganic fillers, as in structural adhesives, the ability to undergo plastic flow in this rubbery region above the glass transition point is significantly reduced. When ultimate elongation is reduced, however, the ultimate stress level is increased and, consequently, ultimate strength becomes well-defined because of an abrupt breaking failure. Thus, when interpreting the failure of joints, the morphology of the adhesive polymeric materials under the conditions leading to joint failure needs to be considered.

In order to develop interfacial adhesive strength in a joint, the polymers present must be involved in wetting, adsorption and interdiffusion interactions with the adherend. When these steps result in adequate joint strength, the stress needed to rupture the joint will be that needed to rupture the adhesive. The stress required, however, is not a well-defined materials constant since testing rate and joint geometry will affect the result even at constant ambient conditions. One can be sure, however, that the maximum survivable stress level will be that of the weakest boundary layer condition in the structure. These weak spot conditions may be due to the overall chemical non-homogeneity of the adhesive materials, or distinct physical flaws may be present. Statistically, we should expect a high probability that somewhere there is a volume element that is much weaker than the average strength of the sample. Thus, with increasing adhesive bondline length or thickness the breaking stress should decrease.

2.2. Adsorption Properties of Polymers

A considerable knowledge of the nature of the adsorption of polymer molecules on adherend surfaces has been developed since the early 1950s. The reader is referred for specific details to the summary of much of this information by Stromberg.[40] Certainly this adsorption of polymers must relate to the durability of the adhesive in joints. Information about adsorption important to the adhesion problem relates to the number and strength of the attachments generated, as well as the availability of other portions of the polymer chain for interaction with other non-attached molecules. Jenckel and Rumbach[41] recognised relatively early that adsorbance of a polymer could correspond to a layer more than ten molecules thick where all segments of the adsorbed chain are attached. Just as no single theory explains all aspects of adhesion, no single theory of the polymer adsorption problem has been generally accepted. Typically, each theory is in agreement with some experimental data but in disagreement with other valid observations. For example, Simha, Frisch and Eirich[42-46] follow the diffusion-equation approach to study the adsorption of polymers. A thermodynamic approach was suggested by Gilliland and Guttoff[47] while Silberberg[48,49] has presented a combination partition-function approach. In general, investigations in this area have been conducted with relatively dilute polymer solutions. Patat *et al.*,[50-53] working with solutions of much higher concentration, showed adsorbances far in excess of earlier investigators, prompting support for the statement that adsorption continues to increase with increasing solution concentration, in the form of multilayer adsorption.

A useful tool was introduced quite recently for the study of molecular adhesion to a metal oxide, namely inelastic electron tunnelling spectroscopy (IETS), which is discussed in detail in Chapter 2. White *et al.*[54] report that IETS provides an incredibly sensitive and versatile method of detecting and identifying molecular species adsorbed on the surface of a metal oxide. It was possible to develop IETS spectra of the multimolecular components of an epoxy adhesive system and of the epoxy mixture adsorbed on aluminium oxide. Comparison by computer calculations with existing infrared optical spectra could then be made. It may be one of the most important methods available for determining the interface physics and chemistry of adhesive bondlines. Its usefulness is also enhanced by the fact that *in situ* studies can be made.

2.3. Rheological Aspects of Polymer Adhesion

Kaelble[55] has published an extensive review on this subject. The molecular properties of polymers that control rheological and adhesion response fall in the categories: (a) chemical composition (monomer units, side groups, secondary bonding forces and cohesive-energy density); (b) molecular structure (tacticity, main-chain symmetry, molecular weight distribution, chain branching, entanglements and crosslinking); and (c) molecular free volume (degree of crystallinity, temperature, pressure and time). Any complete discussion of polymer adhesion must consider the phenomena of bonding, properties of the stable bond and the characteristics of bond fracture. This chapter is intended to focus primarily on joint durability test results for specific polymer adhesive families so more emphasis will be placed on the interfacial unbonding process, which is complicated enough in its own right, than on these more fundamental subject areas (the reader is referred to the Reference list for further details). This complexity is indicated by the disagreement even among prominent investigators as to the possibility of ever encountering true interfacial polymer unbonding. Bikerman[5] contended in 1961, before the development of some of the present surface analysis sophistication, that complete interfacial debonding was not possible. Huntsberger[56] in 1963 demonstrated the regeneration of intact, free surfaces using bonded adherends to prove interfacial debonding did exist. Kaelble[55] interjects the importance of rheology by proposing that the Huntsberger experiment could have shown the opposite result (i.e. cohesive failure) if a regularly increasing temperature or diminished peel rate had been present to change the bulk response. Earlier experiments of Bright[57] had shown such a rate- and temperature-dependent transition between adhesive and cohesive failure where the temperature of peel was controlled. Between the low-temperature branches of the peel curves, which apparently displayed interface failure, and the high-temperature branches, which clearly displayed cohesive failure, was a transition region of combined adhesion–cohesion type failure.

3. MOLECULAR CONTRIBUTIONS TO ADHESIVE PERFORMANCE

Huntsberger[58] has pointed out the need to distinguish between adhesion and adhesive performance. Whereas adhesion may be considered synonymous with the intrinsic interactions across the interface, the adhesive performance is a function influenced by many properties and parameters, such as: gross sample geometry; the topography of the interface; the chemical nature of the materials present; the mechanical responses of the solid; and the viscoelastic phases, strain rates, strain geometry and temperature.

3.1. Molecular Contact at the Interface

It seems logical to accept the premise that good adhesive performance should be present if sufficiently intimate intermolecular contact is achieved to enable London or dispersion forces to exert themselves. Conversely, poor adhesive performance must be associated with limited interfacial contact. Huntsberger[59] has provided supporting evidence for this concept by demonstrating that the temperature dependence of the adhesive performance was markedly influenced by the way in which the adhesive bonds were formed. This behaviour is most satisfactorily explained on the basis of the differences in the extent of interfacial contact at the different temperatures. Kemball[60] has suggested that the mismatch between the atoms in the adherend and adhesive phases can significantly affect the magnitude of the interfacial interactions. He calculated that disparities between the sizes of adsorbing groups and adsorption sites could diminish the shear strength by an order of magnitude. Where atom-for-atom interaction was assumed by Taylor and Rutzler[61] the calculated forces of adhesion were an order of magnitude greater than experimentally measured. Girifalco and Good[12] have introduced a factor which is dependent on both the nature of the interactions within each phase across the interface and the configuration of the molecules at the interface.

If there are specific interactions across the interface—such as hydrogen bonding, ionic bonding or strong dipolar bonding between a high molecular-weight polymer in the adhesive and the adherend—then the average number of segments in contact may approach a direct proportionality with the total number of segments present. Alternatively, a decrease in the number of effective molecular contacts must lead to observations of greater departure from the prediction of a decrease in the force of adhesion with the fourth power of the distance as proposed by Casimer[62] or Lifshitz.[63] It is postulated that if the close contacts were diminished by 50% or more, then the force of attraction between the atoms of the effective contacts would decrease much faster, with increasing separation approaching the seventh power as a limit.

3.2. Wetting at the Interface

The subject of wetting has been widely investigated and discussed in the literature by authors like Zisman,[64-67] Fowkes,[68,69] Good and Girifalco,[70,71] and Johnson and Dettre.[72,73] Sharpe and Schonhorn[74]

J. D. Minford

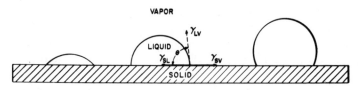

Fig. 2. Wetting and contact angle. θ, contact angle between a drop of liquid and the surface it contacts; γ_{LV}, surface tension at the liquid/vapour boundary; γ_{SL}, surface tension at the liquid/solid boundary; γ_{SV}, surface tension at the vapour/solid boundary.

have supported Zisman in particularly focusing on the role of spreading (see Fig. 2). Study of all the above investigators' contributions, however, shows that conflicting views are held. Sharpe and Schonhorn[74] have contended that the adhesive must exhibit a surface tension at bonding temperature which is less than the critical surface tension of the adherend if a satisfactory adhesive bond is to be achieved. Huntsberger[75] contends that this criterion is not satisfactory because application of thermodynamics has been made to a system which is not representative of any practical adhesive system. He has stated a preference for considering wetting as the process of achieving interfacial contact, and the state of wetting as the number of molecular contacts between the phases comprising the interface as compared with the maximum number of interfacial contacts possible for the system at equilibrium or when wetting is complete.

Obviously, the mixture of materials that comprise most structural adhesives will influence wetting in a more complex fashion than the homogeneous phases used in most scientific investigations. Selective adsorption of one component may provide dramatic changes in wetting rates, or phase separation may occur. In this latter case, the more fluid phase would first wet and fill the interstices on the adherend. If a volatile component is present as in solvent-containing structural adhesive formulations, then the problem is even more complex. For example, a polymer–solvent mixture may pass through its glass transition temperature when the solvent is still present at low concentration. If wetting is incomplete at this time, stresses can build up at the edges of the non-wetted interstices through continuing solvent evaporation. When the localised stresses build sufficiently, a spontaneous separation can occur at the interface or within the adhesive very close to the interface. Consequently, the joint will exhibit very poor adhesive performance in the form of recognisable separation or a greatly diminished bond breaking force. On the other hand, the stresses in an adhesive held above its glass transition point for a longer time while solvent was egressing would be relaxed and wetting by the viscous polymer melt could proceed. Huntsberger[75] has demonstrated that the differences in performance which occur in the same polymer blend through the use of different solvents are a result of the state of the system. If no residual solvent alters the viscoelasticity at equilibrium conditions, he showed the performance would be identical.

Incomplete wetting, for the most part, represents a non-equilibrium state. Maximum wetting rates will be achieved when the capillary pressures are the highest and the viscosity of the adhesive is lowest. Capillary pressures will increase with increasing surface tension of the fluid adhesive. The submicroscopic topography of the adherend also can exert a significant influence on both the capillary pressures promoting wetting and the resistance to the flow required for wetting. The increase in resistance to flow present in the smallest interstices may overshadow a concurrent increase in capillary pressure. It is conjectured that poor adhesive performance can occur even with completely wetted surfaces because the polymer configuration may still prohibit high interfacial contact density. This would also be true if the cohesive strength is particularly low or a relatively low viscosity duplex layer has separated while still wetting the interface.

It is generally accepted that an adhesive used for joining metal surfaces should give strong joints if the adhesive can completely wet the metal surface. In this regard, it has been calculated by Eley and Tabor[76] that physical forces and Van der Waals forces alone are sufficient to give strong adhesive bonds, and the existence of chemical bonds is not necessary to explain strong bonds. Thus, if the metal surfaces were clean, i.e. free of oxide and adsorbed monolayers of various contaminants, then strong chemisorption forces between the metal and the adhesive should be possible. Although an equivalent experiment cannot be cited for aluminium, Baker[77] in 1979 designed an experiment using stainless steel, which has a similar tenaciously adhering thin protective oxide over the metal surface. The metal surface was cleaned by ion bombardment in high vacuum and compared by bonding with an epoxy adhesive with abraded and degreased or abraded and ion bombardment-cleaned surfaces. The ion bombardment-cleaned surface joints proved to be weaker than those from specimens abraded and cleaned in a normal manner. This was presumably because the abrasion produces much rougher surfaces. The strength of abraded and ion bombardment-cleaned joints made in high vacuum, however, was significantly higher and showed 44% higher joint strength retention after soaking in water at 104°F (40°C) for ten weeks.

3.3. The Role of Diffusion

The effects of interdiffusion of the molecules of the adhesive and adherend also need to be mentioned. Voyutskii and Vakula[78-80] have proposed that adhesion between dissimilar polymers as well as 'autohesion' is best explained on the basis of diffusion. They particularly point to the need for explanation for the high adhesion that can be obtained between non-polar polymers or to a decreasing adhesion to a polar adherend which may be observed with increasing polarity of the adhesive. In addition, these investigators point out that electrostatic effects alone fail to provide a satisfactory explanation for many examples of tested adhesive performance. Their evidence was derived from the study of factors such as contact time, time and temperature effect on bonding rate, and the influences of polymer molecular weight and polymer structure; while it is consistent with a diffusion explanation, the data are also consistent with the influences expected from capillarity, rheology and structural effects. In speaking of this evidence Huntsberger[58] has concluded that few adhesive systems will be found for which interdiffusion provides a satisfactory explanation, especially since the polymers present in structural adhesives are usually of relatively high molecular weight, and several investigators[81-83] have shown the incompatibility of most dissimilar high molecular-weight polymers. By its nature, diffusion can occur only across regions where the phases are in contact. Where this contact is possible, the interfacial forces will be sufficiently great to make it inconsequential whether the polymers or polymer segments diffuse across the interface. The performance of the adhesive will most probably be determined by the character and distribution of the non-wetted areas where diffusion cannot even occur.

3.4. Role of Electrostatic Forces

Derjaguin and co-workers[84,85] contend that electrostatic forces alone are both sufficient and necessary to explain observed adhesive performance. Skinner and co-workers[86-88] have demonstrated certain situations where these forces may be significant. Important investigators like Zisman[89] and Bikerman,[5] however, have stated that electrostatic forces are of very little importance. Derjaguin's statement that the action of van der Waal's forces or chemical bonds alone cannot account for an observed high work-of-peeling value is invalidated by his neglect of the dissipated energy, according to Huntsberger.[59] Derjaguin's conclusion, that explanation of the rate dependence of adhesive performance needs consideration of electrostatic forces, also has been invalidated by the quantitative explanation of this effect based on rheology and the time–temperature superposition principle.[58,90–94]

4. JOINT DURABILITY FACTORS

Although an understanding of the intermolecular forces that relate to adhesion and cohesion as summarised by Good[95] is obviously important, we might conclude, when discussing joint durability, that other effects such as bulk deformation or the existence of flaws or residual stresses are of more practical importance. The relationship of such macroscopic properties to intermolecular forces has been the subject of many books.

4.1. Locus of Adhesive Failure

Huntsberger[59] has discussed this matter in some detail. Adhesive joint failure must involve failure of the adhesion in the interfacial area or cohesive failure within one of the joint materials, or some combination thereof. Where separation does not occur at or very near the interface, the problem would not seem to involve adhesion at all. However, this is not entirely true since it has been shown that improper wetting or molecular arrangements of the interfacial layer can specifically influence both the stress concentrations and the cohesive strength of the adhesive that exist near the interface. Consequently, the locus of adhesive failure can be affected even where separation does not occur at the interface.

The problem is very difficult to answer by direct experimentation since residual adhesive films after joint failure may be monolayers or, at most, a few molecular layers thick. Bikerman[5,97] and Sharpe and Schonhorn[74] have stated that interfacial separation is impossible so that only the degree of cohesive failure needs to be considered in failed joints. They draw this conclusion on the basis that the attractive forces involved at the interface are invariably greater than the cohesive strength of the adhesive, and even sometimes the cohesive strength of the adherend. They further state that while apparent adhesion failures are common, they take place so near the interface that the adhesive remaining on the adherend after the bond rupture is not observable. Good[98] has also reported that his analysis of adhesive joint failure led him to conclude that interfacial separation seemed highly improbable, particularly when true chemical wetting of the surface had taken place. Huntsberger[56] has offered experimental verification of interfacial separation in an adhesive/adherend system comprising a low molecular-weight polyisobutylene adhered to an alkyd resin.

More recently, Sharpe[99] has considered the role of weak boundary layers (WBL) in determining the breaking stress of adhesive joints. Weak boundary layers have been a convenient concept for explaining adhesive failure since Bikerman[5] first suggested their existence as a consequence of his assumption that two solids in contact, such as adhesive and adherend, could not fail exactly at the interface. Hence, if failure occurred at or near the interface at relatively low applied stress, a weak boundary layer was probably responsible. There is no doubt that a weak boundary layer is probably an accurate description of the cause of many failures that appear to be adhesive-related. The question remains whether weak boundary layers exist universally, or are there other situations which lead to joint failure and ultimately determine the joint breaking strength? Sharpe[99] looked at this question in a very practical manner: he placed the matter in better perspective by discussing boundary layers generally, including evidence for their existence, their nature where isolatable and identifiable, their development, control of their properties, and their influence on joint behaviour. The reader is referred to the article for details, but a few abstractions can be made which are appropriate to metals. On metals, the intrinsic boundary layers are the oxides whose properties depend on the history of the particular batch of metal and, especially, on the environmental conditions. In this case, the boundary layer properties may be dependent on the conditions used to generate them. The decreasing joint strength accompanying increasing anodising times for aluminium has been shown by Minford[100] to be an example of thicker oxides with more variable composition breaking through planes of weaker structure. The matter of boundary layers in organic polymeric materials is considerably more complex than in metals—primarily due to the wider variety of functional, structural and morphological factors that can be present. The author has found the compositional and structural variability that can occur in aluminium oxides to be sufficiently challenging. Sharpe's main point is that a complex of (interacting) factors enter into the mechanical response of any composite structure under an applied load. As a result, one should not characterise apparently low joint strength as weak boundary layer failure *unless* good evidence exists that a weak boundary layer is really present. It would be more productive to consider the energetics of the deformation and failure process and the detailed mechanics of the joint to explain the low magnitude of the breaking strength.

Acceptance of these ideas, however, offers some complication to acceptance of the theory of adhesive desorption by water. Sterman and Toogood[101] furnished proof of a desorption of the adhesive by water through the use of silane adhesion promoters on bonds to glass exposed in high temperature water. The rate of relative water-diffusion along the interface and through the adhesive was studied by Laird,[102] who found by using glass-reinforced plastic joints that the diffusion of water along the interface could be as much as 450 times faster than through the adhesive. This result may seem difficult to reconcile with the more recent work of Gledhill and Kinloch[103] on the failure mechanism of epoxy/steel joints soaking at various temperatures in water, where water-diffusion rate through the adhesive seemed closely related to loss of joint strength.

Technical papers citing the premature failure of structural bonded metal joints over the years are too numerous to mention them all. Among the earliest reports must be that of Barlow[104] on the Redux adhesive. The pathway for water migration—along the interface or through the adhesive—in these joints would appear to be open for debate, as it was in the Laird[102] or Gledhill and Kinloch[103] studies cited above. The proposed mechanism for environmental failure was a displacement of adhesive by water along the metal, glass or oxide surfaces, respectively. It was concluded by Gledhill and Kinloch,[103] however, that the kinetics of the failure was governed by the rate of diffusion of water through the adhesive to the interface rather than along the interface, as in the Laird study. This variance between the observed preferred water pathway in a metal/adhesive joint as compared with a glass-reinforced plastic joint makes it uncertain whether results can always be extrapolated from studies with different combinations of adherend and adhesive.

Many additional studies have been made in recent years on the effect of water in the bondline. Butt and Cotter[105] have made precise measurements of water absorbed in cured adhesives while Mazor *et*

al.,[106] Farrar and Ashbee[107] and Nicholas and Ashbee[108] have referred
to the observable damage to the adhesive imposed by water in the
environment. Most recently, as described in Chapter 3, Comyn has
cited all the evidence[109] to show the close comparison between decreasing joint strength and overall water uptake by the adhesive.
Evidence is also furnished to show the slow development of microdamage in the adhesive, which is an irreversible degradation process.

Another factor which can disrupt joints between metals and structural adhesives is adherend corrosion. Gledhill and Kinloch[103] established that in water-soaking exposure of steel/epoxy joints, the substrate corrosion did not appear to be operative except as a post-failure
phenomenon. Observations from literally thousands of water-soaking
tests conducted by Minford[100,110-113] indicated that the same conclusion may be valid for aluminium/epoxy joints soaked in water without
significant amounts of chloride ion present, provided that a distinction
is made between the pitting-type corrosion of the aluminium surface
that occurs in the presence of chloride ion-containing waters and the
water-stain corrosion that occurs in non-chloride ion-containing
waters. It has been shown that the debonding process for an aluminium
joint in the latter environment has as one of its elements the conversion of the aluminium oxide interface to aluminium hydroxide, which is
essentially a type of weak boundary layer condition, and is discussed in
detail in Chapters 2 and 5. The acceleration of aluminium joint failure
via an adherend pitting-type corrosion mechanism has been repeatedly
observed by Minford[110-113] where corrosive salt water was present,
such as in a seacoast atmosphere. If the aluminium adherends have
been pretreated to generate a corrosion-resistant surface before bonding, however, such as that afforded by application of certain primers,
anodising or Alodining the surface, then a water-debonding mechanism may be of primary importance in failing the joint.

Krieger[114] has used a salt spray exposure for making comparisons in
durability performance between different chemical types of structural
adhesives and different aluminium alloys. Riel[115] and Rogers[116] were
among the first to describe the possible preferential failure of clad
aluminium joints to bare aluminium joints where corrosive salt water
was in the service environment. Dahringer[117] has developed a test to
show situations where the adhesive itself may cause corrosion of the
adherend and subsequent debonding. Minford[118] has shown that the
metallic oxides added to epoxies to make them electrically conductive
can cause corrosion of aluminium adherends, provided water is present

in the service environment to generate electrolytic corrosion conditions.

Other factors which can influence the adhesive failure are choice of adhesive, bondline thickness and the presence of non-adsorbable or non-desorbable contaminating films at the interface. Choice of adhesive will be discussed later in this chapter when specific durability testing of the different structural adhesive types will be reviewed. The bondline thickness has a significant effect on initial joint strength and durability but for different reasons. One expects—and usually finds—that, in general, increasing the bondline thickness correspondingly reduces initial joint strength. Additional variability in cross-sectional cohesive strength could also result from the variable heat available for curing close to an interface as compared with deeper areas in the bulk of the adhesive. Complex interrelationships exist among the joint separation distance between the adherends, the stiffness of the adherends, and the joint geometry. Also, the presence of a weak boundary layer condition with thick adhesive in the bondline could precipitate an early joint failure. As regards joint durability, however, other factors such as fatigue and crack propagation behaviour of the adhesive or the development of critical bondline flaws as the thickness changes would be very important. An analytical investigation on the influence of bondline thickness upon the stress distribution in single-lap adhesive joints has been made by Ojalvo and Eidenoff.[119] They identified two antisymmetrical adherend/bond interface points at which the shear stresses are sufficiently high to cause the growth of joint failure.

Another important interfacial factor is the presence and relative concentration of non-absorbable or non-desorbable contaminating films on the adherend, as shown in Fig. 3. Whilst it has been mandatory to ensure that these kinds of contaminants are not present for aerospace-type bonding applications, they are commonly present in many general-manufacturing bonding operations, including such major fields as the motor industry. In the fabrication of metals, rolling lubricants are commonly present whilst special mould-release agents may be required in forming plastic sheet or moulded products. In assembling metal structures by adhesive bonding, assembly of the parts is often preceded by certain metal-forming operations, which result in surface contamination by forming lubricants. Best bonding durability can only be expected after a separate operation to remove the lubricant prior to adhesive application. Minford[120] in 1974 showed that for

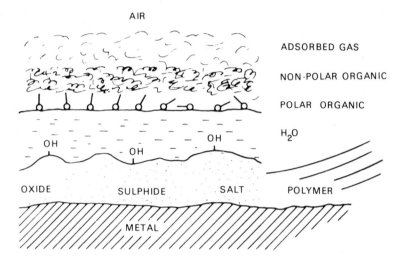

Fig. 3. *Hierarchy of spontaneously adsorbed layers on a metal surface.*

any joints made in the presence of surface contaminants, lower joint durability should be expected, compared with joints made using specially treated adherends. However, Minford[121,122] has recently shown that significant progress has been made since 1980 in formulating paste-type adhesives containing epoxy, acrylic and vinyl plastisol resins which have amazing tolerances for oil contaminants on metal adherend surfaces. Danforth and Sunderland[123] have deliberately controlled surface contamination and determined the effect on the newest 121°C (250°F) curing epoxy structural film adhesives.

5.   ADHESIVE FACTORS AFFECTING JOINT DURABILITY

5.1. Modulus of Elasticity

The high modulus of a cured adhesive in the bondline coupled with a significant difference in the coefficient of thermal expansion between the adherends can be important factors in causing early joint debonding. Where large differences do exist, it is usually beneficial to use an adhesive of low modulus, such as a nitrile rubber–phenolic adhesive, as a stress relief interlayer. The best adhesive choice must be a compromise between adequate strength to support the structure and the ability to withstand the brutal, and generally alternating, stresses which strain the bondline under varying temperature service conditions.

5.2. Interfacial Imperfections

DeBruyne[124] and Bascom[125] have furnished examples where an imperfection such as a trapped air bubble in the cured adhesive can be the site for high localised stress, which can accelerate the debonding effect of any desorbing agent in the service environment, such as water (see Fig. 4).

Some degree of interfacial imperfection can arise even through the normal adhesive-curing process. For example, Plueddemann[126] has listed the adhesive shrinkage which can accompany curing as a principal factor for bond failure. Adhesives which cure after wetting the surface of the adherend by solvent evaporation, cooling or chemical polymerisation mechanisms are all candidates for such failure.

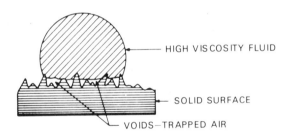

HIGH VISCOSITY FLUID

SOLID SURFACE

VOIDS—TRAPPED AIR

LOW AREA OF INTERFACIAL CONTACT RESULTING FROM HIGH VISCOSITY OF FLUID

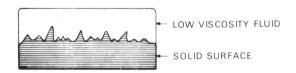

LOW VISCOSITY FLUID

SOLID SURFACE

LACK OF VOIDS AND HIGH AREA OF INTERFACIAL CONTACT RESULTING FROM LOW VISCOSITY OF FLUID

Fig. 4. Effect of adhesive viscosity on interfacial contact.

Yet a different situation can develop if the bond interface is altered due to adhesive swelling. Admittedly, this is a rarer occurrence for structural thermoset adhesives than for thermoplastic-type adhesives, but all adhesives probably can be affected to some degree depending on the type of fluid present. Metal adherends constitute a particularly sensitive situation since they neither absorb any solvents nor change in volume to compensate for any volume change in the adhesive.

5.3. Heat Curing

The performance of bonded joints can be positively affected in several ways by the heat cure. At elevated temperature, the lowered adhesive viscosity can more readily lead to better wetting and spread on the surface. Even with some soiling present, better bond durability may be achieved because the lower-viscosity adhesive can absorb, dissolve, disperse or desorb the soil. This is clearly a factor in the observation by Minford[127] that it is virtually impossible to make strong or durable aluminium joints using heavily oiled aluminium and two-part paste, room temperature-curing epoxy adhesives; yet, both strong and durable joints can be made with highly contaminated surfaces and certain new heat-curing epoxy pastes.[121,122] Minford[128] has found that short exposures to higher service temperatures can alter the adhesion of two-part epoxys previously cured for long times at room temperature on aluminium adherends. Failures originally appearing adhesive-type are converted to 100% cohesive-type, and the joint durabilities are greatly enhanced.

Sell[129] has also shown that the higher the temperature of the cure, the greater the tendency to find greater joint durability. Although 177°C (350°F) curing epoxys have been replaced in many aerospace applications with 121°C (250°F) curing systems, there is little disagreement that the higher temperature curing systems are generally the more durable in aggressive water-soaking exposure conditions. It has even been shown that a 121°C (250°F) curing system that has been purposely cured at 177°C (350°F) will show an increase in durability. It is difficult to prove, however, the relative influence of such factors as changes in modulus or the number and/or type of attachments to the substrate. Sell[129] has also shown that within a given family of adhesives the joint durability will increase as the creep resistance increases in the series. This might favour the modulus-increase argument to some degree.

5.4. Pressure

Pressure on the adhesive in the bondline can also have a positive effect on durability in several ways. It can promote better wetting and spreading of the adhesive when applied in conjunction with heating. It can be the physical factor for forcing adhesive into a surface of marked roughness or porosity. It can help reduce interfacial imperfections like air bubbles or voids and increase uniformity in the bondline. Finally, in the curing of any structural adhesives whose polymers can form volatile products, the use of relatively high, constant pressure on the bondline is mandatory. Vinyl–phenolic adhesives, for example, can suffer significant reductions in what is otherwise good durability performance if the recommended curing pressures are not maintained.

5.5. Fillers

There are several technical papers in the literature on the effect of fillers on durability in epoxy adhesives. Bodnar and Wegman[130] have shown the superior resistance to various natural weathering conditions of filled polyamide/epoxy bonded aluminium joints as compared with unfilled Epon 828/Versamid 140 formulation joints. The superior resistance of filled epoxy formulations as compared with unfilled Epon 815/Versamid 125 blends has also been shown by Minford[111,131] in various accelerated laboratory water-soaking tests of aluminium joints. A possible distinction between inorganic fillers was also shown by Minford in which epoxys filled with aluminium powder or a china clay produced more durable joints than an unidentified mineral filler in a proprietary commercial-adhesive product. This latter epoxy adhesive formed the only epoxy-bonded aluminium joints (including two unfilled epoxy varieties) which failed within 8 years in an industrial atmospheric exposure. To what degree this poorer performance could be directly associated with the use of the particular filler could not be demonstrated because the corresponding unfilled joints were not available. The fact that aluminium powder as a filler in Epon/Versamid mixtures could produce aluminium joints which lasted twice as long in Minford's[131] corrosive seacoast exposure tests certainly seems significant.

The effect on joint durability of aluminium powder filler in nitrile-modified heat-curing epoxys like 3M's EC-2086 (with filler) and EC-2186 (unfilled) has been studied by Bodnar and Wegman[130] and Minford.[131] The effect of the filler in the atmospheric weathering exposures of Bodnar and Wegman was inconclusive since both types of joints failed at about the same time in a jungle exposure; filled epoxy joints were superior in the desert but poorer in an industrial atmosphere. Similarly, there was no superiority shown by either the unfilled or filled adhesive joints in water-soaking tests by Minford. Spathis *et al.*[132] have recently suggested a model describing the influence of spherical exclusions in epoxys. They concluded that a number of desirable effects might result, such as stiffening of the matrix and an increase in the average strength properties of the polymer. The adhesion of filler to matrix was found to depend strongly on the diameter of the particles. Of much less significance was the volume content of the filler.

5.6. Adhesive Thickness

Wolfe *et al.*[133] found that aluminium joints with thicker bondlines had noticeably shorter fatigue lives, a not unexpected result after earlier comments about the locus of failure in thick bondline joints. Wake[134,135] has developed the argument that the weakness of thicker joints is due to stresses built into the joints by the contraction of the adhesive on setting or by differential contraction between adhesive and adherend after curing at elevated temperature. This implies a linear relationship between strength and thickness.

Bryant and Dukes[136-138] made many measurements of the strengths of joints over a wide range of bondline thicknesses. In their earliest work they confirmed that an effect exists, in general indicating the thinner the joint, the stronger the bond. Later, using four different joint designs and room temperature-curing rubber or tough epoxy adhesives, they concluded that their data were best described by relating either log strength or linear strength to log thickness. The dependence of joint strength on adhesive thickness seemed mostly a statistical consequence of rupture process. They also introduced the concept that the variations in bond strengths with thickness was in part due to the change in the rate of straining which occurs with increased thickness.

5.7. Temperature

Foulkes *et al.*[139] have studied the nature of the joint failing process associated with its temperature change. Even though thermosetting resins in adhesives have no softening point, they still show a change in joint strength at higher temperatures that seems to be quite independent of any thermal degradation process. Foulkes offered evidence that the adhesive fails by a brittle mode at low temperature but, at room temperature, many adhesives begin to fail by a ductile mode. Thus, the recorded bond strength decrease with increasing temperature is probably due to a plastic flow phenomenon.

Because temperature is a common factor in all service environments, its effect on bond permanence should be considered. Service temperature ranges of primary interest from about −55°C (−67°F) to 93°C (200°F) during World War II have risen in recent years to −253°C (−423°F) to 538°C (1000°F). Generally speaking, the problems reflecting on bond permanence with varying temperatures are the result of stress concentrations and gradients developed within the joint. These stress concentrations are due to many causes, including: differences in thermal coefficient between adhesive and adherends; shrinkage of adhesive in curing; trapped gases or volatiles evolved during bonding; modulus of elasticity and shear strength differences between adhesive and adherends; differences in thermal conductivity; and residual stresses in the adherends as a result of release in bonding pressure. Operating at the lower cryogenic temperatures intensifies stress concentrations because of the magnified differences in the physical and mechanical properties of adhesives and adherends. For example, a low-modulus adhesive which can stress-relieve itself at room temperature by deformation will increase in modulus of elasticity at cryogenic temperatures to the degree that stress-relieving is no longer possible. Increasing the area of both adherend and joint can multiply these stress situations. In some bonded structures on the space shuttle, for example, the sensitivity of certain joints to debonding was not demonstrated until a full-size module was joined and immersed in a cryogenic liquid nitrogen bath. All testing with scaled-down test specimens had failed to bring a potential problem into focus.

Black and Blomquist[140–143] were among the first to try to elucidate the causes of joint failure at temperatures up to 260°C (500°F). They found a significant interaction between the surfaces of different metals and adhesives of the phenolic–epoxy type which they were formulating for use in high-temperature service applications. Stainless steel surfaces degraded the adhesion of the same adhesive that was unaffected by aluminium adherends at 260°C (500°F). It was discovered that the adhesion to aluminium surfaces was also degraded if iron or copper was deposited on the aluminium as a pretreatment. Overall, it was found that bond permanence and strength at test temperatures up to 316°C (600°F) were significantly affected by varying the ratio of the resins, the acid accelerators present, and surface stabilisers such as the chelating agents used to pretreat surfaces.

More recent work on developing high-temperature-resistant polyimide adhesives with greater fracture toughness by St Clair and St Clair[144] deserves study.

Kausen, in a state-of-the-art review article[303] in 1964 on adhesives for high and low temperatures, compared the best anticipated bond performance based on the heat resistance and thermal ageing properties of the adhesive classes available at that time. A summary of his work can be seen in Fig. 5. For best permanence under cryogenic conditions the various adhesive families have the relationship shown in Fig. 6.

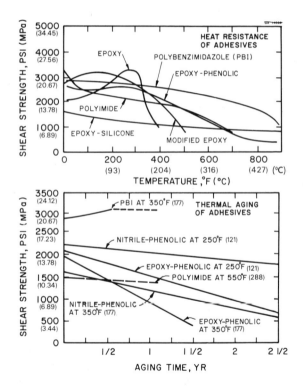

Fig. 5. Comparison of high-temperature structural adhesives: the polyaromatics (PBI and PI) have the highest shear strength at temperatures above 800°F (427°C) (top) and best retention of strength on ageing (bottom).

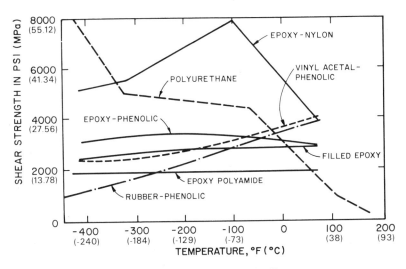

Fig. 6. Properties of cryogenic structural adhesive systems.

5.8. Adhesive Fracture Mechanics

It is probable that the adhesive joints in real structures are always blemished by flaws of various types and sizes, i.e. bubbles, dust particles or unbonded areas (see Fig. 4). It is the progressive separation radiating from these flaws that basically controls the strength of the bonded structure.

The fracture-mechanics approach to evaluating joint performance is related to the early work by Griffith[145] in 1920 using brittle materials like glass. Irwin[146] later used the concept to measure the resistance of a material to fast crack extension in the presence of a flaw.

It was Ripling and co-workers in 1963, however, who applied this background to evaluating adhesive responses in joints. They concluded that the overriding factor in establishing the toughness of any bonded joint is the speed with which the crack moves along the bondline. Low toughness will be characterised by abrupt jumps of the crack while slow-moving cracks predict a high degree of bond permanence, i.e. toughness. Of the many technical papers forthcoming from Ripling and Mostovoy since 1963, only a few are cited here on adhesive subjects such as measuring plane-strain fracture toughness, stress-environmental cracking, fracturing characteristics and joint geometry effects.[147–157] The most characteristic specimen used to evaluate crack

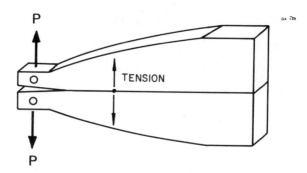

Fig. 7. *Bonded tapered double cantilever beam* (TDCB), *used for quantitative evaluations of fracture toughness and environmental durability effects on bonded systems.*

propagation effects is shown in Fig. 7. The Ripling and Mostovoy work with this tapered double-cantilever beam (TDCB) and a simpler form designated the uniform double-cantilever beam (DCB) led to a still simpler adaptation of the concept by Bethune and co-workers at Boeing[158–161] in the form of a thin-adherend, uniform DCB specimen better known as the 'wedge test' (Fig. 8). Much attention has been focused on this test for evaluating the relative durability potential of different bonding systems based on a measurement of the relative rate of crack growth during exposure to high humidity at elevated temperatures. Two experiments tried at Boeing are exemplary. Sections of actual aircraft were removed where disbonding was present and compared with other sections where no disbonding had been evident. When wedges were inserted in the bondline and the structure was exposed to 100% R.H. at 60°C (140°F), complete fracture of the bondlines occurred in less than an hour for those sections where disbonding had been noticed earlier. The sections free from disbonding in service when exposed to the same conditions remained intact for many hours.

In a second experiment, surfaces to be bonded were treated in

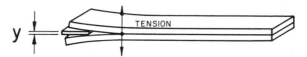

Fig. 8. *Thin adherend, uniform DCB* ('wedge test'), *not used for quantitative evaluations because of possible plastic deformation of adherends.*

sulphuric acid/sodium dichromate etching tanks. The crack propagation rate was studied as the conditions of the tank were varied. The necessity to maintain certain dissolved levels of aluminium and copper were established for treating aluminium aircraft in order to obtain optimum bond durability values. This procedure has been equally applicable to studying the effect of time and temperature in etching and even the effect of rinse delay on acid-etched adherends. Boeing has used the criterion in manufacturing of accepting only bonding conditions which lead to a crack propagation rate of less than 1·9 mm/h (0·075 in/h) in a 100% R.H. at 48·9°C (120°F) exposure condition.

It is clear from the above that the variable performances with the wedge test procedure have been responses to changes in the surface pretreatment of the adherend, and it is accepted today that the wedge test results for predicting joint durability are primarily a way of distinguishing between surface preparations rather than between adhesives. The test can, however, be used to predict relative durability of different adhesives as well, by using a known good-performance surface pretreatment as a constant bonding parameter with a variety of different adhesive candidates as the only experimental bonding parameter.

An example of the fracture-mechanics approach for studying differences among adhesives can be shown from a Mostovoy and Ripling[162] investigation using the TDCB-type specimen. Six proprietary heat-curing structural adhesives were investigated. As a group, the adhesives all showed the general behaviour of having a high resistance to monotonically increasing loads (G_{Ic}) over the test temperature range from $-17\cdot8°C$ (0·F) to 93·3°C (200°F). At room temperature and below none of the adhesives seemed to be sensitive to water under sustained loads, but above 37·8°C (100°F) their resistance to water under static loading decreased rapidly. At 65·6°C (150°F) only the nitrile–phenolic adhesive cured at 121°C (250°F) showed good flaw tolerance; i.e., G_{Ic} (in H_2O) values more than 1 in lb/in^2 (0·18 kJ/m^2). The decrease in flaw growth resistance was particularly pronounced for the nylon–epoxy system which was already known to have poor durability in bonded joints exposed to water. Loads above this G_{Ic} (in H_2O) value result in cracking rates that are excessive, and catastrophic failures result in relatively short times. In metals, the fatigue growth rates, even above a threshold value, are sufficiently slow for metal structures still to be designed for finite service life. Adhesives, however, generally crack rather rapidly above the threshold value so that finite service life design does not appear to be feasible. It was apparent from the work that the critical material property for adhesives is their resistance to fatigue crack growth, particularly if the structure is expected to experience an alternating load.

Bascom and co-workers[163] have used yet a different special combined-mode adhesive joint specimen—a scarf-type joint—to study the development of micro-cracks and their propagation and interaction with main crack failures. The locus of bond failure in such joints was in the adhesive resin but less than 1000 Å from the interface. This near-interfacial failure was judged to be the result of the stresses directing the failure into the interfacial area. If this is true, then apparent interfacial failures of this type should be expected for many joint designs when the loading results in both tensile and shear stresses acting on the adhesive.

The most comprehensive and recent summary of the fracture mechanics approach to joint failure has been given by Kinloch and Shaw.[164]

As has been mentioned in section 4.1, much evidence has been offered that apparent interfacial failures are, in reality, cohesive since a thin film of the adhesive is left on the substrate. Jennings,[165] Mulville[166] and Wilcox and Jemian[167] all achieved such failures using butt joints loaded in tension, butt joints loaded in shear and thick, bent lap-shear specimens, respectively.

Jones and Kaelble[168,169] have conducted investigations directed toward the determination of the fundamental material parameters that are descriptive of fracture behaviour of adhesive joints. Methods were developed to measure physical-mechanical and fracture behaviour of adhesive joints under carefully controlled environmental conditions at several temperatures. The high temperature-resistant adhesives did not display transitions in mechanical properties at room temperature and, thus, did not display crack propagation in fatigue experiments conducted at room temperature. However, the crack growth rates could be altered by exposing the adhesive bonds to various chemicals which apparently could compete favourably with the adhesive by adsorption for the various interfaces in the joint. Other chemicals could diffuse into the adhesive and plasticise the adhesive at the crack tip. Stable crack growth was found in fatigue testing in each test medium, and at low cycle rates the test medium effect would decrease. It was proposed that the corrosion effects of water at the adherend interface also could increase the fatigue crack propagation rates in long-term fatigue.

The results of more than ten years work in the Naval Research Laboratory to develop an adhesive fracture test methodology and data base were summarised by Bascom *et al.*[170] in 1978. Their work demonstrated the more highly complex nature of fracture testing of structural adhesive joints than monolithic structures. Especially complicating is a very strong bond thickness effect and the need to perform the testing under combined stress conditions. The authors concluded that major work remained to be done on developing the analytical relationships for the effect of joint geometry on adhesive fracture behaviour.

The development and extension of a crack is so critical in joint durability evaluations that some efforts to establish the feasibility of automatic detection and classifications of bondline defects in adhesively bonded structures should be mentioned. Loew *et al.*[171] have been working in this area using multiple-layer bonded structures. Detection of a bond flaw was expanded to include its classification as an unbond, void or porosity condition. The evaluation was a scan with a $0 \cdot 5$ in transducer in the pulse–echo mode. Flaws could be found with accuracies of 91% as to their presence and 92% as identifying the type of flaw. A so-called false-alarm rate of 19% was found where no identifiable defects existed. In only 1% of the situations, however, was a defect called a non-defect. The procedure seemed to work regardless of the layer depth of the flaw.

Clark[172] has worked with specimens with intentionally induced adhesive-bondline flaws of the type designed and built into the PABST program, where the highest present state of bonding technology was employed. His tests showed that bondline flaws loaded primarily in shear did not grow, while those loaded by peel or tension forces did exhibit flaw growth. Also, flaw growth rate was higher at the slow rate of 2 cycles/h in fatigue testing than at the faster rate of 30 Hz.

Cuthrell[173] has studied the change in micro-cracks in adhesives with increasing load and concluded that growth was exponential, while Zhurkov *et al.*[174] have looked for the sources of crack initiation in epoxy structural adhesives.

Closely related to the study of joint durability and micro-cracks, voids and potential propagation to failure of defects in the bondline, are the many investigations to detect defects in bondlines using non-destructive test procedures. Of the voluminous literature available in this area, only a few investigations will be identified as representative. On ultrasonic adhesive bond evaluation, there have been relatively recent technical papers by Schliekelmann,[175] Matzkanin *et al.*[176] and Dukes and Kinloch.[177] Arnold and Vincent[178] were certainly among the first to attempt a better understanding of possible correlations between ultrasonic reflection parameters and bond quality. Rose and Meyer[179] in 1973 and Rose[180] in 1977 have shown significant progress in fine tuning the process to predict and classify different levels of bond strength. Tattersall[181] has referred to observations which he interprets as defining a so-called 'slackness' at the bond interface relating to poorer potential joint durability. Neutron radiography, which has been used for assessing the structural serviceability of the adhesive in the joint by Dance and Peterson,[182] is an example of the many new techniques investigated in recent years.

Reference should be made, when considering flaws in bondlines, to the investigators who have attempted to predict joint durability by taking constant stress data and combining them with reaction rate theory as applied to polymer mechanical behaviour.[183,184] McAbee and Levi have shown that such treatment is applicable to some adhesive data under constant rate of loading conditions. Different reaction-rate equations were developed for three decreasing relative humidity situations. Under 90–95% R.H. exposure conditions, the equations were relatively successful in predicting failure times for AF-126 film adhesive/aluminium joints. Under 50% R.H. conditions the data scatter was considerably greater. At 20% R.H. the derived equations were not useful. It was reasoned that the lower humidities do not promote significant weakening of the bonds and, hence, random flaws in the bondlines probably play a more prominent role in inducing bond failure.

Another matter related to the adhesive failure contribution to adhesive joint durability is whether the stress involved in the service environment is constant or intermittent. Certainly with aircraft and most vehicle structures loads are applied and relaxed at intervals. There has therefore been some trend in durability testing of aerospace bonds towards procedures in which the load is repetitiously applied and relaxed. This cyclic stressing, in turn, should be applied under environmental conditions consistent with weathering factors in the service environment. Some of the more comprehensive studies in this regard have been undertaken by Frazier and Lajoie[185] at Bell Helicopter. The selection of a load–unload period is critical because adhesives vary in the rates at which they creep under load and relax when the load is reduced.

The principle used at Bell can be demonstrated by comparing a relatively low temperature-curing adhesive like FM-123-7 with 177°C (350°F) curing nylon–epoxy FM-1000. The FM-123-7 adhesive polymer shows a yield point in its stress-strain curve but no work-hardening properties. Creep-under-stress measurements showed the almost instantaneous deflection was 0·0069 cm (0·0027 in), in the next 15 min it increased by 0·0008 cm (0·0003 in), and after the next 45 min it had risen by 0·00038 cm (0·00015 in), When the load was relaxed, all displacement relaxation was completed within 2 min. For this adhesive a load time of 30 min and unload time of 5 min would seem to be a desirable cycle. An extremely tough, high-elongation adhesive like FM-1000 shows quite different creep and relaxation properties, and a different cycle would be desirable. After 1 h the FM-1000 polymer is still creeping, although at a slower rate. Initial deflection was similar to the FM-123-7, i.e. 0·0061 cm (0·0024 in), but within 30 min, additional creep was 0·0071 cm (0·0028 in), and after 60 min, an additional 0·0020 cm (0·0008 in) of creep was still measured. The same sequence occurred in reverse when the load was relaxed, but relaxation is still measurable after 15 min. Based on these observations it was decided that a maximum bond damage cycle for the FM-1000 joint would probably be 2 h of loading followed by 1 h of relaxation.

Whilst such differences in adhesive behaviour can be anticipated and measured, economic considerations generally preclude such measurements. Bell eventually had to settle for evaluation at a compromise load cycle for all other adhesives under evaluation of 27·56 MPa (4000 psi) for 1 h and unload for 15 min. Testing at only one load cycle condition also precludes development of enough data to construct a curve of durability versus stress level.

Interfacial crack propagation studies have also been conducted by Mulville and Vaishnav;[186] the effect on fracture toughness was measured for four different surface finishes on aluminium surfaces bonded with epoxy adhesive. Both tensile and bending load tests were conducted with polished, milled, glass-peened or sandblasted aluminium adherends. It was found that increased surface roughness resulted in greater fracture toughness with the interfacial cracks replicating the surface features of the aluminium. The crack propagation occurred in the epoxy near the interface, with the residue of epoxy remaining bonded to the aluminium. An additional important finding was that the residual stresses due to casting and curing of the epoxy could contribute 15–20% of the strain required for crack initiation at the interface.

In all the studies mentioned so far, the properties relating to propagation of defects were inherent in newly cured adhesive. Grimes[187] has studied the effects of long-term adhesive storage on the fatigue resistance of structural adhesives.

5.9. Adhesive Stress Endurance

It is generally acknowledged today that the most discriminating means of determining joint durability is to impose simultaneously stress and weathering conditions on the bondline. Marceau and Scardino[188] assessed the various durability test methods in 1975 as part of a comprehensive metal bond durability programme. Most recently McMillan[189] has thoroughly reviewed the durability test methods for aerospace bonding.

Before a more detailed discussion of stress-endurance testing of longer duration, however, it may be pertinent to note the results for tests where the stress is raised continuously to joint failure with the bondline kept wet and in a peeling mode. Carter[190] introduced this technique, which is best described as a wet peel test. He considered it was most suitable for evaluating short-time environmental effects but probably not suitable for long-term durability prediction. Recent testing by Locke[191] at Boeing has been reported to have pointed out situations where this test was more discriminating than the Boeing wedge test.

A variety of means of imposing a steady stress on an adhesive joint have been suggested over the years. These include Eickner's procedure[192-194] shown in Fig. 9. In these tests, a steady stress was maintained in a qualitative, thorough, accurately described manner. The dramatic bond failures shown in the later work of Sharpe was not as evident in these early tests, indicating that the stress levels were probably not very high in the Eickner investigation. A more quantitative method of imposing and holding a sustained stress was proposed

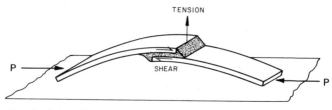

Fig. 9. Standard lap shear in bending test. Specimen 2·54 cm (1 in.) wide, or entire assembly, is bent and time to failure recorded.

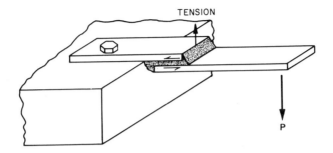

Fig. 10. Cleavage lap shear test. Time to failure is recorded.

by Sharpe[195] using a calibrated spring jig. The use of the spring-loaded jig does guarantee that a known stress is being imposed and that the stress will not be relieved by accommodations in the bondline. Some early investigators also used a cleavage lap-shear test specimen as shown in Fig. 10. Whilst this kind of specimen may be capable of distinguishing between different adhesives, the values probably have little practical significance because a lap joint would never be included in a structure to resist such severe cleavage forces. Carter[196] was an early investigator in this area who imposed stress by simply bolting groups of lap joints together and attaching weighed concrete blocks to the bottom joint. The 3M-type stressing fixture can be considered a blend of these ideas, where the individual blister-detection type joints (Fig. 11) are bolted together inside a jacket where hot humid weathering conditions can be maintained.[197] The applied stress is via a spring-loading device attached externally to the jacketed chamber (Fig. 12). A stressing fixture introduced by Alcoa (as shown in Fig. 13) has been used extensively since about 1970 by Minford.[100,110–113,198,199] While

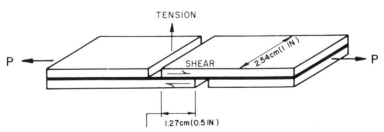

Fig. 11. Blister-detection lap shear test (MMM-A-132). Time to failure is recorded.

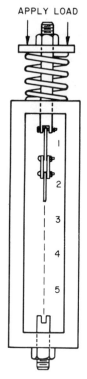

Fig. 12. 3M's spring-loaded fixture.

there is no clear understanding of all the mechanisms involved in stress–environmental type failures, the work of Bascom[200] nicely summarises the current state of understanding with an appropriate list of references.[200–208]

The data obtained with these various stressing procedures are usually plotted as a stress–endurance curve, seeking to predict at what level of stress the adhesive in the joint could be expected to have an indefinitely long service life (see Fig. 14). A stress–endurance plot by Minford[100] using the Alcoa stressing fixture to compare the effect of different surface pretreatments on aluminium shows typical data in Fig. 15.

Using high resolution microscopy, Venables et al.[209] suggest that resin deformation and fracture can be a considerable part of the

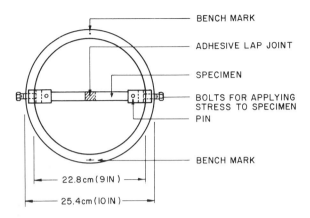

Fig. 13. Alcoa stressing fixture details.

adhesive stress–environmental cracking process. Clearly, there is a need to select the oxide morphology, primer and adhesive combination that can develop an optimised interphase to offer good stress–environmental cracking resistance.

5.10. The Adhesive in a Water Environment
Water must be considered a dominant factor in determining the permanence of adhesive-bonded joints in most weathering service environments. In this process the adhesive may be affected or involved in a variety of ways, depending on whether water can permeate the

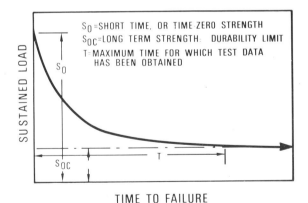

Fig. 14. Durability limit curve in shear mode.

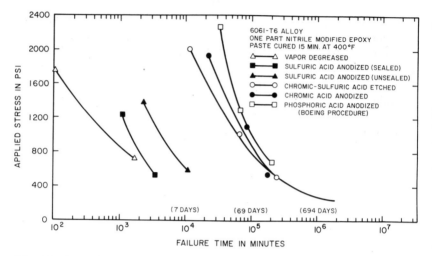

Fig. 15. *Joint durability of vapour-degreased, chromic–sulphuric acid etched and various anodised surface joints stressed and exposed in 100% R.H. at 125°F (52°C).*

adhesive structure itself to produce significant property changes or move via some pathway to reach the interfacial area and affect adhesion.

Bascom[210] has published several papers on how water can degrade the strength of various materials, including both the adhesive and the adherend. He believes fracture mechanics provides a fundamental and quantitative means of describing material failure by various environmental factors including water. While failure usually begins with the formation of one or more minute flaws or cracks at the interface, it is the application of stress at a magnified level at the crack tip which begins the propagation. The action of an adsorbate like water must be to alter the bond strength at the crack tip, i.e. effect a reduction in the cohesive energy. An alternative mechanism may be that the water embrittles the material around the crack tip, reducing the amount of deformation that can occur ahead of the crack. Under conditions where the crack is propagating rapidly this may be an unlikely mechanism because it would require a very rapid diffusion of water to the crack tip and into the deformed area. Moisture might also assist failure by creating surface corrosion of the adherend, which can lead to surface flaws such as pits or inclusions of a weak corrosion product.

Also, if water can condense into a crack tip, it should be capable of exerting considerable capillary pressure, thus acting to open the flaw. Bascom reminds us that significant condensation does not usually occur except at relative humidities of 90% or greater. Since most accelerated laboratory weathering tests are conducted in 95% or higher humidities, these mechanisms are of significance; however, they say nothing about the atomic or molecular processes involved.

Some of the earliest investigations on the unique effect of water on joint durability were by Falconer et al.,[211] with further confirmation by other investigators.[212-214] The experiments of Kerr, MacDonald and Orman[213] comparing the relative effect of dry ethanol or water on a cured epoxy polymer (whether inside or outside the bond joint area) furnished specific evidence for the preferential action of water at the interface as compared with action on the bulk adhesive (see Fig. 16). While alcohol was more readily absorbed into the polymer matrix, reducing the tensile strength by 30%, there was no bond strength deterioration in the corresponding aluminium/epoxy joint after soaking

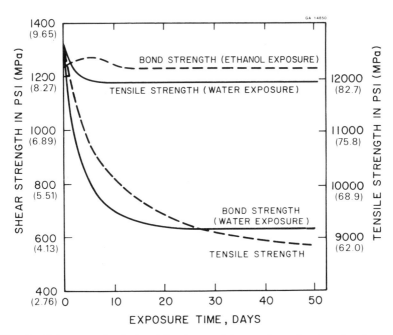

Fig. 16. *Joint strength of aluminium/epoxy joints and tensile strength of epoxy resin after exposure to water or ethanol at 90°C (194°F).*

in alcohol for the same period. By contrast, water in the environment was barely absorbed at all by the cured epoxy and the epoxy tensile strength was lowered by only 8%, yet the adhesive joint strength was lowered by 50%. Orman and Kerr[215] developed typical epoxy/aluminium bond strength decay curves over a 50-day exposure period in 100% R.H. at 40°C (140°F) or 100°C (212°F) as shown in Fig. 17.

As expected, the higher-temperature humidity condition degraded the joint strength more significantly than at lower temperature, but the time to reach a lower limiting value was about ten days at either temperature. The adverse effect of the water on joint strength proved to be somewhat reversible when the weakened joints were exposed to heating at 90°C (194°F) in vacuum for varying times as shown in Fig. 18.

More recently, Brockmann[216] has studied the change of strength and deformability of the adhesive layer itself under hot and humid climate conditions, and concluded that the structural adhesive part of the joint is involved in a nearly reversible process and can reach a predictable equilibrium condition. However, Brockmann concludes that the de-

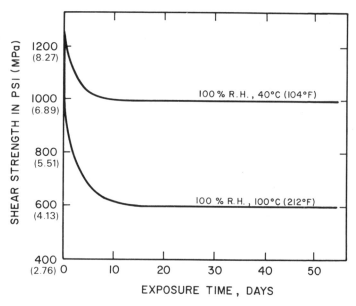

Fig. 17. *Typical joint strength decay curves of aluminium/epoxy bonds.*

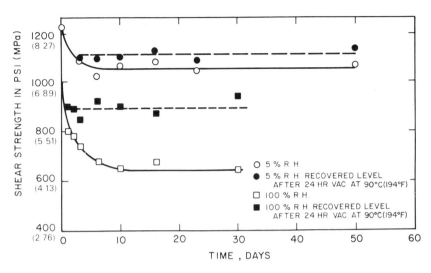

Fig. 18. Exposure of aluminium/epoxy joints at 90°C (194°F).

terioration of the actual adhesive forces in long-term aged joints is an irreversible process proceeding in an unpredictable manner toward catastrophic destruction. In the case of aluminium bonds, Brockmann found the water stability of the boundary layer in the joint was dependent both on the chemical structure of the adhesive and on the type of surface to be bonded. It was pointed out that two of the most effective ways to increase the stability of the boundary layer to humidity were to change the adherend surface condition through special surface preparations, and to use corrosion-resistant primers. The progress made by using these techniques, however, has been achieved without an exact knowledge of the nature of the adhesion forces existing between the organic adhesives and the adherend. We can project that the adhesion between structural adhesives and most metal adherends used for joining is due to physical and, possibly, chemical bonds. After wetting the relatively rough metallic adherend surface in its low molecular-weight and reactive form the adhesive may build up chemical bonds with the oxides and hydroxides on the surface. In addition to the chemical linkages such as phenolates or alcoholic bondings with phenolic resins or epoxide resins, there are a large number of hydrogen bonds that can occur in the boundary layer.

The special question now becomes, in what manner does this adsorbed resin layer react in the water environment? In experiments where polyester or phenolic resins were adsorbed on aluminium oxide, it was shown that up to 50 and 25%, respectively, of the total resin could be desorbed after soaking in pure water. This would indicate that while the bonds between structural adhesives and metals may be chemical in nature, they are not absolutely water-stable. Of course, there is no reason to expect that phenolic, alcoholic or hydrogen bonds should not deteriorate in the presence of water. In joints with large bonding areas and long diffusion paths, however, the effect of water should be greatly delayed compared with the described experimental water-soaking conditions above. One might consider how it would be possible to replace the chemically stable, but water-unstable, bonds with more water-stable chemical linkages. Brockmann[217] has demonstrated one possibility by using a chelating agent like 8-hydroxyquinoline or 5,7,2,4-tetrahydroxyflavonene (Morin) which can react with aluminium oxide in alcoholic solutions to form water-stable aluminium complexes. When phenolic resins or nitrile epoxies were cured on these converted surfaces, there was a significant increase in joint strength retention following water-soaking exposures.

Gledhill and Kinloch[103] worked on identifying the mechanisms and kinetics of joint failures due to water and successfully predicted the instability of joints from thermodynamic considerations. The mechanism of failure for the mild steel/epoxy joints was one of displacement of the adhesive on the ferric oxide surface by the water. A similar type of failure would be expected for any epoxy-to-metal oxide substrate surface bond. A further elaboration of this investigation by Kinloch[218] and Gledhill, Kinloch and Shaw[219] separated the process into three discrete steps: (i) displacement of the adhesive on the oxide governed by the rate of diffusion of water through the adhesive to the interface; (ii) loss of strength and failure of the oxide due to subtle changes in its nature; (iii) under the special conditions where corrosive water is present, there can be gross corrosion of the substrate as a failure mechanism. This mechanism would indicate that more stable oxides and stronger interfacial forces must be forged from better surface preparations and adhesive selections to resist rupture by water. These investigators, using steel/epoxy joints, also showed experimentally that a critical water concentration was reached for the debonding mechanism to operate. When this level is exceeded, there will be a relatively sharp boundary between the outer weakened regions of the joint and the inner regions which are not yet affected by the environmental attack. The extent of the debonding at this point will be equivalent to an environmental crack length which may be calculated and subsequently used to predict a value for the fracture stress under a given set of time, temperature, and water-soaking conditions. Similar conclusions might be anticipated for aluminium/epoxy joints since the kinetics of the displacement of the adhesive from the aluminium oxide surface may also be governed by the rate of diffusion of water through the adhesive to the interface.

6. STRUCTURAL ADHESIVE FAMILIES AND RELATED DURABILITY DATA

6.1. Structural Adhesive Categories

Structural adhesives are most conveniently discussed by referring to their physical form in use rather than their chemical composition. As such, they are first described as being solids, liquids or spreadable pastes.

The solid adhesives group, in turn, may be broken down into the film adhesives (unsupported), tape adhesives (supported) and solid powders or preforms.

A second broad-category group may be referred to as '100% solids' paste and liquid adhesives, which may subsequently be subdivided into one-component (long shelf life) and two-component (short pot life) groups. The term 100% solids can be somewhat misleading, however, as small amounts of volatile solvents can be added for special reasons. Small amounts of solvent may improve adhesion to some contaminated metal surface or, in the case of plastics, provide some 'surface bite' to improve adhesion. Structural adhesive primers are usually dilute solutions of a high molecular-weight resin which may be the same as or chemically compatible with the 100% solids structural adhesive to be used to make the joint. Sometimes the reason for adding a little solvent to a high molecular-weight resin may be simply to increase spreadability.

Finally, there are some structural adhesives available which are truly solvent-based because the weight ratio of solvents to resin is greater than unity. It is vital to the development of strong, durable joints to allow all of the solvent to evaporate in these adhesives before heat curing.

The one-component group may be heat-cured or cured by surface or anaerobic catalysis while the two-component group is usually cured at room temperature, but properties and service durability will generally be improved with a shorter elevated-temperature cure.

Most important in structural bonding applications are the tape and film solid adhesives; the 100% solids, one-component, variously catalysed liquids and pastes; and the two-component, room temperature-cured pastes.

6.2. Chemical Polymer Families

Whilst blends of thermoplastic polymers and elastomer polymers can be present in structural adhesives to alter significantly the physical properties and response of the joint to stress and the environment, the chemistry of structural adhesives is essentially the chemistry of the thermoset epoxy and phenolic resins. Bolger[33] and Claret[21] discuss the chemistry of these resins in sufficient detail for most purposes.

Table 1 lists alphabetically the structural adhesives mentioned in this chapter.

6.3. Phenolic Resins in Structural Adhesives

Although two types of phenolic resins exist (resol and Novolac) structural adhesives usually contain the alkylated (methylated or butylated methylol group) resol variety. Such resins are compatible with epoxy and other polymeric materials used in formulating structural adhesives. The resol phenolics possess exceptional wetting and spreading ability on metal and metal oxide surfaces due to the high concentration of strongly adsorbed phenolic and aliphatic hydroxyl groups present. Their extreme brittleness and high shrinkage during curing, however, prohibits their separate use as structural adhesives. Various fillers and modifying resins are necessary to reduce the shrinkage and provide stress relief. The main modifying resins most frequently co-reacted with resol phenolics are linear, high molecular-weight polymers with recurring hydroxyl groups along the polymer backbone. Reaction then proceeds between these hydroxyl groups and the phenolic methylol group. Examples would be epoxy–phenolic and vinyl–phenolic adhesives.

It is rather easy to envisage a reaction between the phenolic resin and the epoxide linkage, but the vinyl designation in the vinyl–phenolic adhesive description is somewhat misleading. However, the actual

TABLE 1
STRUCTURAL ADHESIVES

Commercial name	Manufacturer	Chemical type
ADX 41.2	Hysol	Modified epoxy
ADX 6562	Hysol	Modified epoxy
Aerobond 3030	Hysol	Modified epoxy
AF-30	3M Company	Nitrile phenolic
AF-55	3M Company	Polyether–modified epoxy
AF-126	3M Company	Nitrile epoxy
AF-126-2	3M Company	Nitrile epoxy
BR34/FM34	American Cyanamid	Polyimide
BR127	American Cyanamid	Nitrile phenolic
EA 9628	Hysol	Modified epoxy
EC-2086	3M Company	Nitrile epoxy
EC-2214	3M Company	Nitrile epoxy
Epon 422	Shell Chemical	Epoxy phenolic
Epon 815	Shell Chemical	Bisphenol A–epoxy
Epon 828	Shell Chemical	Bisphenol A–epoxy
FM-47	American Cyanamid	Vinyl phenolic
FM-61	American Cyanamid	Elastomer epoxy
FM-73	American Cyanamid	Elastomer epoxy
FM-96	American Cyanamid	Vinyl epoxy
FM-123-2	American Cyanamid	Elastomer epoxy
FM-123-7	American Cyanamid	Elastomer epoxy
FM-400	American Cyanamid	Modified filled epoxy
FM-1000	American Cyanamid	Nylon epoxy
HT-424	American Cyanamid	Epoxy phenolic
Metlbond 227	Narmco Materials	Elastomer epoxy
Metlbond 302	Narmco Materials	Epoxy phenolic
Metlbond 324	Narmco Materials	Modified epoxy
Metlbond 1113	Narmco Materials	Modified epoxy
Metlbond 1133	Narmco Materials	Modified epoxy
Narmco 105	Narmco Materials	Vinyl phenolic
Plastilock 677	B. F. Goodrich	Modified epoxy
Plastilock 717B	B. F. Goodrich	Modified epoxy
Redux 775	Ciba–Geigy Group	Vinyl phenolic
Superbonder 324	Loctite Corp.	Anaerobic acrylic
Superbonder 325	Loctite Corp.	Anaerobic acrylic
Versamid 125	Henkel adhesives	Polyamide
Versamid 140	Henkel adhesives	Polyamide
XA-3408	3M Company	Nitrile epoxy

resins used are poly(vinyl formal) (PVF) or poly(vinyl butyral) (PVB) resins which can average as many as 46 hydroxyl groups per chain.

6.3.1. Urea–Formaldehyde and Phenol–Formaldehyde Adhesives

When choosing from this general class of adhesives, it is necessary to consider the adherend since different choices would be made according to different adherend properties. For wood bonding and plywood laminations, the urea–formaldehyde and phenol–formaldehyde adhesives have been most commonly accepted (see Chapter 8). Their lower cost, easier handling, good durability in moisture, and ability to penetrate into wood structures recommend their use over epoxys. In contrast, they are too brittle for use with metals, glasses and other rigid impermeable adherends.

6.3.2. Vinyl Phenolics

The 'Redux' adhesive system of the Aero Corporation used for joining De Havilland aircraft in World War II was the earliest successful structural adhesive of this class. A resol–phenolic solution is spread on one adherend and a powdered poly(vinyl formal) resin is sprinkled over the top. The joint is closed after the solvent has evaporated and heat and pressure effect a cure. This toughening of a phenolic resin through co-curing with a high molecular-weight linear polymer led to hitherto unheralded higher bond strengths and durability for structural joints. Although these resins will crosslink with heat alone, the reactions are significantly accelerated with small amounts of strong acid.

As predicted, higher phenolic ratios will increase the temperature resistance at the expense of some flexibility and peel strength. Substituting PVF for PVB resins increases both hot strength and shear strength. Many demonstrations of the high reliability and durability of this adhesive in metal joints have been offered over the last 30 years. Solomon[220] gives many references citing excellent durability. Long-term durability testing has been conducted by Minford[221] using 6061-T6 aluminium and two proprietary vinyl phenolics (FM-47 and Narmco 105). The Narmco 105 etched-surface joints survived two years exposure to room temperature water soaking, 100% condensing humidity soaking at 52°C (125°F), or a challenging soak/freeze/thaw cycle condition for two years with no significant change in joint strength. The bond strength declines for the FM-47 joints were moderate to highly significant in the same exposures, indicating that testing of one member of a chemical class of adhesives does not necessarily reflect the performance of all other members. In a four-year industrial atmospheric exposure of these same two adhesive types of joints, there was no distinction in performance with each showing no strength shifts. In corrosive seacoast exposure after four years, the Narmco 105 joints again proved superior in durability with no change in joint strength recorded. Figure 19 from the data of Minford[111] illustrates how variations in choice of adhesive family can alter the durability of etched aluminium joints in the corrosive seacoast exposure.

The vinyl–phenolic structural adhesives have been especially important in the development of bonded aircraft structures in Europe. Catchpole[222] reviewed some of these application areas in 1965. A review of the excellent durability performance of these adhesives in Fokker aircraft was issued by Koetsier[223] in 1975. DeLollis[224] in a 1977 review of the published stressed and non-stressed durability data of Olson *et al.*[225] reported the good durability of both vinyl–phenolic Redux 775 and FM-47 formulations. Hockney[226] has also furnished evidence of the good performance of vinyl–phenolic adhesives under natural weathering conditions.

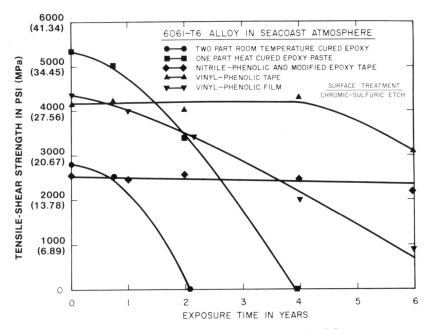

Fig. 19. *Effect of adhesive variation on joint durability.*

Other durability testing involving the evaluation of vinyl–phenolic structural adhesives has been conducted by Rogers,[227] Eickner and Schowalter[228,229] and Novelli.[230]

In spite of this excellent durability record, there has been a continuing concentration on developing newer types of structural adhesives which can provide lower curing pressures and temperatures and higher peel strengths, at both low and elevated temperatures.

6.3.3. Nitrile Phenolics

These very durable types of structural adhesive are made by blending nitrile rubber and a phenolic Novolac resin. This is the exception to the statement at the beginning of section 6.3 that resol–phenolic resins are primarily used in formulating structural adhesives. Long-chain alkane groups instead of reactive methylol groups must be present in a phenolic resin to be compatible with the nitrile rubber. This class of materials has accounted for the largest poundage volume of tape, film or solution structural adhesives over many years. The exceptional durability of nitrile–phenolic bonded metal joints has been evaluated

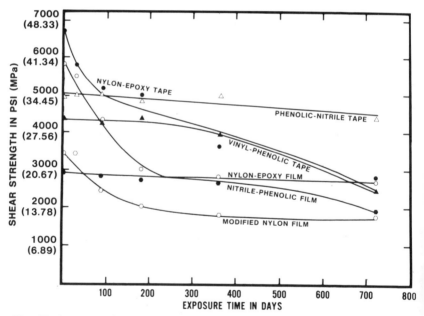

Fig. 20. *Tape and film structural adhesives, soaked in water at room temperature.*

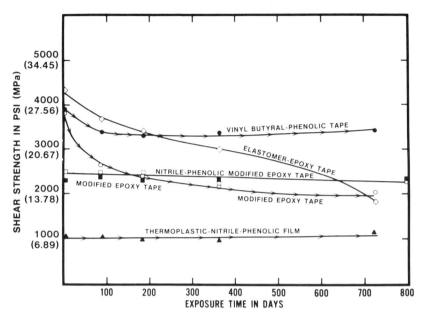

Fig. 21. Tape and film structural adhesives exposed to condensing humidity at 52°C (125°F).

by many investigators. Minford[111] has recorded the excellent performance of several nitrile–phenolic adhesives, as in Figs. 20 and 21, where the joints were soaked for two years in room temperature water or condensing humidity at 52°C (125°F). Similar excellent durability was found during four-year exposures to an industrial or a seacoast atmosphere, as shown in Figs. 22 and 23.

Many investigators have reported the benefit of adding carboxyl groups to nitrile phenolics for increasing adhesion to metals.[231–233]

Of particular interest is the amazing ability of nitrile phenolics to form bonds to aluminium that have outstanding resistance to undercutting corrosion even in corrosive salt water. This has been one basis for using dilute solutions of nitrile phenolics as primers for structural bonding systems.

Bethune[234] has produced data to demonstrate the satisfactory service of nitrile–phenolic adhesives in aircraft structures during more than 25 000 h. By contrast, room temperature-curing epoxy joints had failed in a few thousand hours, and even 121°C (250°F) curing

J. D. Minford

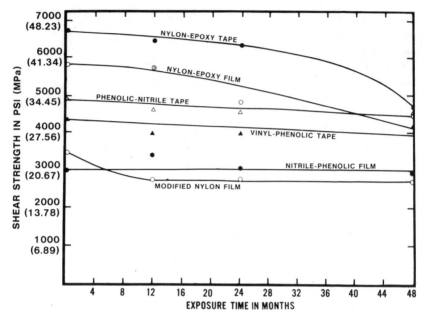

Fig. 22. *Tape and film structural adhesives, exposed to an industrial atmosphere.*

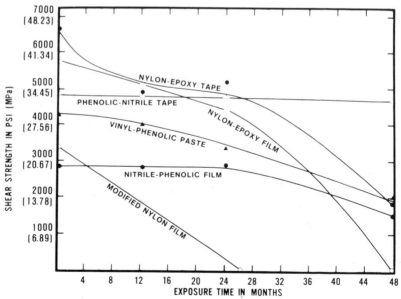

Fig. 23. *Tape and film structural adhesives, exposed to a marine environment.*

modified-epoxy systems had shown sporadic disbonding. All these adhesives had met the US Military Specification MMM-A-132 used for acceptance at that time. This need for better durability discrimination eventually led to the development of the Boeing wedge test specimen as shown in Fig. 8. Very recently, Wolfe *et al.*[235] have demonstrated the superior fatigue resistance of a nitrile–phenolic to a nitrile–epoxy adhesive. They concluded that an adhesive with lower modulus of elasticity and higher strain-to-fracture value would be expected to withstand better oscillating peel stresses.

Bodnar and Wegman[130] found AF-30 nitrile–phenolic aluminium joints completely resisted deterioration from exposure in Yuma (Arizona), Picatinny Arsenal at Dover (New Jersey) and in the jungles of Panama.

6.3.4. Epoxy Phenolics

Epoxy–phenolic type adhesives were primarily developed in the 1950s in an effort to improve the long-term thermal stability of the commonly employed vinyl–phenolic structural adhesives. Adding phenolic resins to the acetals had certainly improved their overall thermal stability, but the acetal portions still tended to decompose with long service above 112°C (250°F). Black and Blomquist[140–142] at the Forest Products Laboratory were the first to combine newly available, high molecular-weight epoxy resins with resol phenolics in generating the new adhesive class of epoxy phenolics. As a substitute for the PVB or PVF resins, the epoxy backbone polymer offers a major improvement in thermal stability but with the expected sacrifice in flexibility, elongation and peel properties. Typical commercial adhesives like Epon 422, Metlbond 302 and HT 424 generally show cured shear strengths on metals in the 13·78–20·67 MPa (2000–3000 psi) range with good strength to 260°C (500°F). Bodnar and Wegman[130] evaluated HT-424 epoxy–phenolic adhesive bonded aluminium joints in three atmospheric conditions, and in three years the joints sustained higher losses than those fabricated with a simple unfilled polyamide–epoxy system like Epon 828/Versamid 140. Of particular concern was a loss of 22% in strength in Yuma, Arizona, where the polyamide–epoxy joints suffered no losses.

Minford[236] has obtained an apparently higher degree of bond durability for an epoxy phenolic than the Bodnar and Wegman data indicate above. Epon 422 joints made with etched Alclad 2024-T3 aluminium retained 81% of their initial strength after three years continuous soaking in 63°C (145°F) water.

The good high-temperature resistance of HT 424 and Epon 422 epoxy–phenolic joints has been reported by Kuno.[237]

6.4. Epoxy Resins in Structural Adhesives

There are several excellent texts on the subject of epoxys.[238-240] The most important single class of epoxy resin is that derived from bisphenol A. A liquid bisphenol A with an epoxide equivalent of 300 or less can be used to formulate most one-part or two-part paste or liquid structural adhesives. Higher epoxide equivalent resins (400 or higher) are used in the tape- and film-type structural adhesives. The lower epoxide-equivalent epoxy resins cure by addition reactions involving the terminal oxirane group and a second compound with an active hydrogen. Some of this same cure mechanism may also occur in the solid epoxy resins (higher epoxide-equivalent resins) used for tape and film adhesive manufacture, but cure more frequently proceeds through the hydroxyl group for these epoxy adhesives.

6.4.1. Room Temperature-Curing Epoxy Adhesives

These adhesives usually consist of a liquid epoxy resin (component A) plus an aliphatic amine or polyamide (component B). To control the viscosity and colour and to reduce shrinkage in thickness, pigments or fillers like aluminium, silica or carbonate must be added. If the viscosity is too high, certain low molecular-weight diluents such as butyl or phenyl glycidyl ethers can be added. The simple aliphatic polyamines will effectively cure the liquid epoxys but, because of their skin- and respiratory-irritant action, are less frequently used. In addition, the high resin-to-catalyst ratios, short pot life and rigid, fracture-prone nature of such adhesive systems are objectionable. More commonly used as hardening agents are the polyamides, which are reaction products between polyamines and organic acids. A desirable added toughness may result from polyamide curing but often at the expense of some water resistance. The most convenient commercial form of room temperature-curing epoxy has been the equal parts of A and B variety. The data from Minford[111] in Figs. 24 and 25 demonstrate that a variety of unfilled and filled room temperature-curing epoxy formulations afforded aluminium joints of excellent durability in long-term water-soaking exposures when the aluminium adherends were chromic–sulphuric acid-etched. In Fig. 26 correspondingly good performance for all but one epoxy was shown after four years of exposure to an industrial atmosphere.

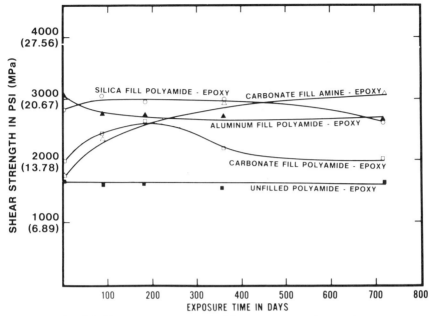

Fig. 24. *Two-part epoxys, soaked in water at room temperature.*

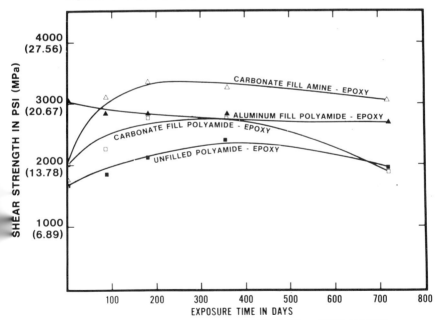

Fig. 25. *Two-part epoxys, exposed to 100% R.H. at 52°C (125°F).*

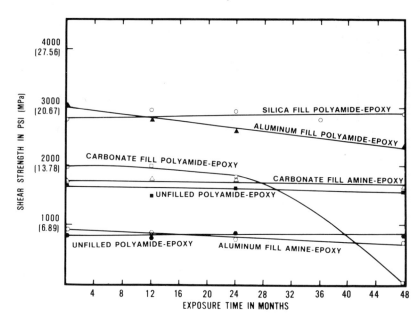

Fig. 26. Two-part epoxys, exposed to an industrial atmosphere.

Eickner,[241-243] as early as 1955, claimed good performance for polyamide–epoxy joints in the Panama jungle with clad aluminium. Later, he[194] reported some stressing tests of these epoxy joints in the jungle using the method shown in Fig. 9. Most recently, Minford[244] has reported very satisfactory 12-year exposure results for polyamide–epoxy etched aluminium joints in both open and secluded jungle exposures in Surinam. The two-part room temperature-curing epoxy joints actually performed more durably than heat-cured nitrile–epoxy joints. Bodnar and Wegman[245] similarly showed a preference for the polyamide–epoxy over the nitrile–epoxy in exposures in the Panama jungle. Cotter[246] obtained best performance in the jungle from epoxy–Novolac and nitrile–phenolic formulations with a polyamide–epoxy adhesive showing poor weathering resistance. The comparative evaluation by Minford[111] of a variety of two-part epoxy for two years in an aggressive soak/freeze/thaw weathering cycle showed there can be considerable variability in durability among commercial room temperature-curing epoxys.

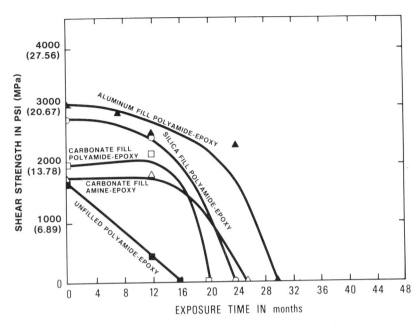

Fig. 27. Two-part epoxys, exposed to a marine atmosphere.

In Fig. 27 Minford[111] has shown the poor bond durability of the same group of room temperature-curing epoxys in the corrosive sea-coast atmosphere. This shortened joint life results from undercutting corrosion of the aluminium adherend even though chromic–sulphuric acid-etched adherends were used.

It should be remembered, however, that durability performance of room temperature-curing epoxys can be altered by varying other bonding parameters such as surface preparation, stress level and filler selection, and by elevating the curing temperature. Most investigators rightly indicate that room temperature-curing epoxys generally have a special need for surface preparation on metal adherends. However, the recent data of Minford in a jungle environment showed remarkably high joint-strength retention after 12 years exposure using such an epoxy and only vapour-degreased Alclad 2024-T3 adherends. Durability of comparable vapour-degreased 6061-T6 joints was noticeably inferior. Remarkable improvement in the durability response of aluminium pretreated in different ways and bonded with a two-part room temperature-curing epoxy has been demonstrated by Minford.[112] The very rapid failure of such bonded joints when an exposure including both significant stress and water-soaking elements are employed has been demonstrated by Sharpe[195] and Minford[110,111] as shown in Figs. 28 and 29.

Schlies[247] has published data on the long-term storage of two-part epoxy joints at ambient temperature, and Olson[225] has exposed stressed and non-stressed two-part epoxy joints in natural weathering, as has Cotter.[246] Levi[248,249] has used a hot water soak reaction-rate method for evaluating the durability of two-part epoxy joints which would see stressing and humidity in service. At Picatinny Arsenal a wide ranging programme to evaluate the durability of various adhesive families has included many two-part epoxys. Some of the results have been reported by Tanner[250] in 1965 and again in 1972.[251] Wangsness[252] has reported the results of 3M's effort to establish the sustained load durability of their structural adhesive product line, which included several room temperature-curing two-part epoxys.

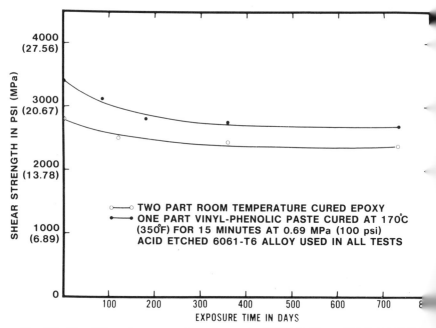

Fig. 28. *Durability of unstressed specimens in 100% R.H. at 52°C (125°F). Compare with Fig. 29.*

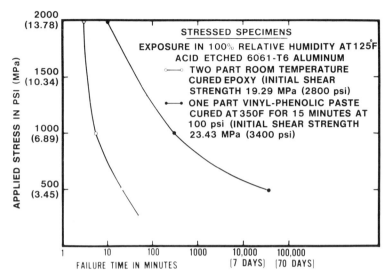

Fig. 29. Durability of stressed joints.

6.4.2. Elastomer Epoxys

The trend in recent years has been towards toughening epoxys with elastomers and choosing catalyst systems which enable curing, typically using dicyandiamide (dicy), at relatively low pressures and temperatures in the shortest possible times. These curing properties have been very helpful for the assembly of ever larger wing, tail and fuselage sections in aircraft, permitting lower steam pressures and thinner-walled autoclaves. The dramatic improvement effect on tensile shear and T-peel of adding nitrile rubber to an epoxy can be seen in data from Lewis and Saxon.[238]

A number of authors[253–257] have discussed the effect of the addition of a liquid rubber such as carboxyl-terminated polybutadiene/acrylonitrile (CTBN) elastomer to epoxy resins. The fracture energy can be increased 30–40 fold. In the additive range to 15%, the elastomer precipitates and cures as spherical particles which can be seen in the fracture surfaces of the epoxy. Above 15%, the CTBN forms a more or less homogeneous blend, and the fracture energy declines sharply. Of particular interest is the fact that the gain in toughness occurs with little loss in tensile strength, modulus or thermal mechanical resistance of the epoxy matrix resin. Bascom *et* al.[258] have described the role of the elastomer particle as permitting a much larger volume for plastic deformation at the crack tip than is found in the unmodified resin.

The durability data of Minford[111] for several nitrile–epoxy adhesives (XA-3408, EC-2086, EC-2214) on etched aluminium are shown in Figs. 30 and 31 for hot humidity and hot soak/freeze/thaw exposure cycles for two years. It is clear that the nitrile-rubber modification imparts a distinct sensitivity to hot water-type service conditions. In an industrial atmospheric exposure for four years (see Fig. 32) the same adhesives formed quite durable joints, but all similar joints had failed before four years in a corrosive seacoast exposure, as shown in Fig. 33. As compared with the performance of two-part epoxy joints in the same atmosphere (see Fig. 27) the nitrile–epoxy joints were more durable overall. The author attributes some of this better performance to the better surface wetting achieved when the one-part nitrile–epoxy is heat-cured at 177°C (350°F).

Bodnar and Wegman[245] have also evaluated EC-2086 nitrile-epoxy/aluminium joints in the desert, industrial and jungle environments, and all joints failed between two and three years in the jungle.

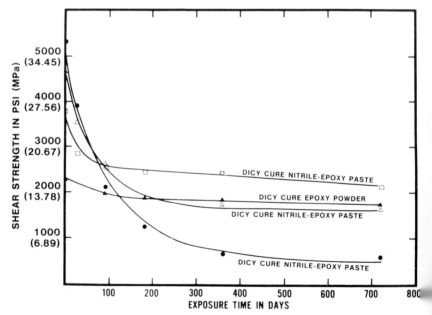

Fig. 30. *One-part epoxys exposed to 100% R.H. at 52°C (125°F).*

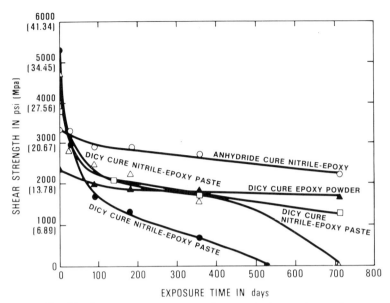

Fig. 31. *One-part epoxys, exposed to a water immersion cycle.*

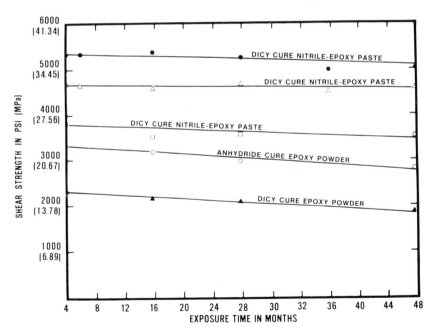

Fig. 32. *One-part epoxys, exposed to an industrial atmosphere.*

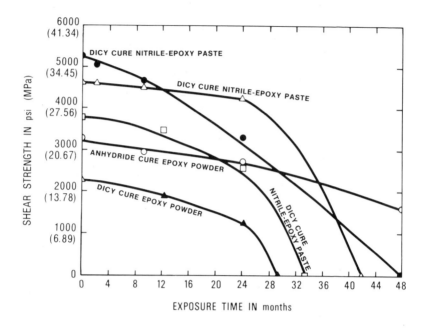

Fig. 33. One-part epoxys, exposed to a marine environment.

Even in the dry desert atmosphere the loss in joint strength for
EC-2086 and EC-2186 adhesive joints averaged 15%, indicating a
significantly lower durability than that of corresponding vinyl– or
nitrile–phenolic joints.

Wangsness[252] has reported the sustained load durability of the 3M
commercial nitrile–epoxy paste and film adhesives. Hughes and
Rutherford[259] have tested Metlbond 227 (nitrile–epoxy) along with
aluminium-filled EC-2214 nitrile–epoxy under stress and high humidity conditions whilst Grimes[260] has looked at the tension–tension
fatigue performance of AF-126-2 nitrile–epoxy joints.

Schwartz[261] has compared the stressed durability performance of a
177°C (350°F) curing nitrile–epoxy with two 121°C (250°F) curing
modified-epoxy film adhesives with varying adherend surface preparation interaction. DeLollis[262] has evaluated both FM-123-2 and EC-
2214 nitrile–epoxy joints in continuous immersion in water and high
humidity.

6.4.3. Nylon Epoxys

The availability of non-crystalline nylons that are soluble in alcohols provided the means of blending nylon with epoxy resins to yield curable structural adhesives. These adhesives yielded unusual increases in peel, shear, fatigue and impact compared with adhesives developed earlier. However, these adhesives have never been able to circumvent several serious drawbacks such as loss in peel strength at low temperature, poor creep resistance and extreme sensitivity to moisture. DeLollis[263] has documented this sensitivity to moisture by offering comparative aluminium joint data with a conventional nitrile–phenolic film adhesive. While the nitrile–phenolic joints lost only 16% strength after 18 months in high humidity exposure, the nylon–epoxy joints degraded 80% in two months. Minford[111,264] has shown data (Fig. 34) where nylon–epoxy tape/aluminium joints with initial strengths above 41·34 MPa (6000 psi) failed in a few days in condensing humidity when stressed to a small fraction of their initial failure loadings. Many of the efforts to offset this moisture sensitivity involved the use of resol–phenolic resin-based primers. While these primers did prolong service life they did not raise the performance levels to those of nitrile rubber or acetal-toughened epoxy structural adhesives. In Fig. 35 two-year exposure of unstressed FM-1000 nylon–epoxy joints is shown for room

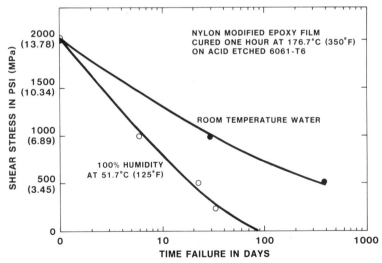

Fig. 34. *Durability of stressed nylon-modified epoxy/aluminium alloy joints.*

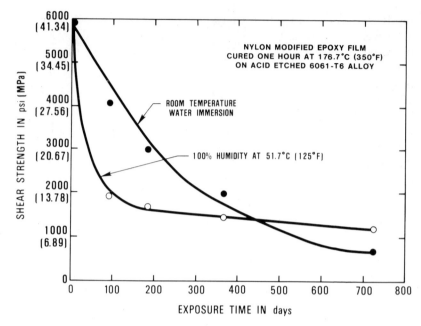

Fig. 35. *Durability of unstressed nylon-modified epoxy aluminium alloy joints.*

temperature water immersion or 100% condensing humidity at 52°C (125°F). Whilst joint strength declined by 67% in the first 100 days in hot humidity, this level was not reached until 400 days in lower temperature water. Some sort of equilibrium condition seemed to be achieved after 100 days in hot humidity, however, so the final strengths of the joints after 700 days were very similar.

Minford[198,199] has studied the effect of varying stress levels on the joints made to a wide variety of aluminium surface pretreated joints with FM-1000 nylon–epoxy film adhesive. Different weathering conditions such as condensing humidity at 52°C (125°F) or alternate immersion cycles in corrosive salt water were simultaneously imposed along with the stressing.

Kuno[237] has studied the effects of higher temperatures on Metlbond 406 nylon–epoxy joints.

6.4.4. Modified Epoxys
For a considerable number of commercial structural adhesives, modifying resins used are not specified in the manufacturer's literature. As a

result, the consumer has no basis for conclusions about potential adhesive/joint durability other than test data reported by the manufacturer, published in the literature or developed in his own operations.

In conducting various durability tests of a particular duration with proprietary tape and film structural adhesives, Minford[221] found six of 12 commercial products were only described by the formulators as being modified-epoxy types. The adhesives so described included Metlbonds 324 and 329, FM-61 (one side of duplex tape was modified epoxy with a nitrile–phenolic as second side), FM-96, Plastilock 677 and Aerobond 3030. The joints fabricated were evaluated in room temperature water, hot humidity, hot soak/freeze/thaw cycle, and industrial and seacoast atmospheric conditions. The variation in durability performance was so wide ranging among these adhesives that no significant conclusion about service potential could be attached to their common description of being a modified-epoxy structural adhesive. For example, four categories of relative durability performance were used to describe the results: (i) joints could increase in strength; (ii) they might show no significant change; (iii) the joints decreased moderately in strength; (iv) there was highly significant loss in strength. After two years in room temperature water four modified epoxy-type formulation joints of those tested were in categories (i) and (ii). Two of the adhesives, however, shared moderate losses in strength and were therefore in category (iii).

In a more aggressive hot soaking humidity exposure at 52°C (125°F), there were three adhesive candidates in the top two durability categories and two in the moderate joint strength loss category. One commercial formulation showed significant weakened joints after two years. In an aggressive soak/freeze/thaw weathering cycle which readily discriminates between many commercial adhesives only two modified epoxy tape and film adhesives were unaffected. A single adhesive was in the moderate-loss category while two other adhesives showed significantly deteriorated joints. Even after four years in a milder industrial atmospheric-weathering condition, there was wide variation in performance. Three adhesives formed joints with the expected performance of no change in joint strength but there was a decline in strength of joints fabricated with two of the proprietary products. In the corrosive seacoast environment all the modified-epoxy adhesives were unaffected even after four years and so fell in category (ii). It is interesting to note that the most uniform durability performances were observed in the last, most corrosive type of weathering.

Perterka[265] has shown the good performance of paint top-coated, heat-cured epoxy joints after ten years exposure to a central European climate. Hughes and Rutherford[259] evaluated the influence of stress and weathering on modified epoxy ADX-41.2. Wangsness[252] included five modified-epoxy 121°C (250°F) curing adhesives in his sustained load testing. Schwartz[261] used both stressed blister-type lap joints and wedge specimens to evaluate two 121°C (250°F) curing modified epoxy formulations. Cotter[246] included a modified-epoxy heat-curing adhesive in his high-humidity jungle exposures for comparison with room temperature two-part and nitrile–epoxy joints. Grimes[260] used Metlbond 1113 modified-epoxy to compare in constant cycle fatigue with both a nylon–epoxy and a nitrile–epoxy structural adhesive.

Askins and Schwartz[266] have evaluated modified-epoxy adhesives like FM 73 and FM 400 for metal sandwich panel bonding to resist accelerated adverse weathering. They[267] have also evaluated the durability of FM 400 adhesive for joining graphite/epoxy facings to aluminium honeycomb cores. Shannon and Thrall[268] in the PABST (Primary Adhesively Bonded Structure Technology) programme have evaluated four candidate modified-epoxy 121°C (250°F) curing adhesives (M1133, AF55, FM 73 and EA 9628) with BR 127 corrosion-inhibiting primer under varying stresses and weathering conditions. McMillan[269] has looked for the best relationships between surface preparations on aluminium and 121°C (250°F) curing modified-epoxy adhesives. Wegman[270] has conducted both stress durability and fatigue tests using four 121°C (250°F) curing modified epoxys (ADX 6562, EA 9628, M1113 and PL 717B). Finally, Scardino and Marceau[271] have conducted extensive wedge-type testing of commercially available 121°C (250°F) curing modified-epoxy adhesives.

6.4.5. Testing of Primers

It has been accepted in the aerospace industry that the quality and assurance required for structural joints in that industry mandates the use of special adhesive primers. Since a primer used in structural joints must essentially permit the joint strength to approach the cohesive strength of the adhesive, the primers used must be of commensurate high strength. Primers must be able to flow over the surface and, therefore, they are generally diluted structural adhesive formulations. In recent years, the additional requirement that the primer should have corrosion-inhibiting characteristics has been encouraged. The best review of new primer technology in recent years has been presented by

Reinhardt,[272] based largely on experimental work performed at the Air Force Materials Laboratory at Dayton, Ohio.

The materials and procedures at present used to evaluate primers for structural bonds are described in the technical literature. For example, Noland[273] has conducted extensive testing of primers with various heat-cured epoxy formulations in high humidity exposure conditions. Falcone and Miller[274] have considered the benefit of primers in regard to making repairs on main rotor blades in helicopters. Spencer[275] has investigated the role of primers in bonding both aircraft and space structures, and Aker[276] also has studied the function of adhesive primers in bonding aircraft structures.

Data on the effect of primers on durability are included by Shannon and Thrall,[268] Schwartz[261] and Scardino and Marceau.[271]

7. STRUCTURAL ACRYLICS

These adhesives are actually two-part systems like room temperature-curing epoxys, but they may or may not have pot-life problems depending on whether the activator is mixed into the resin or applied as a separate primer-like solution. They consist of a solution of acrylic polymers and monomers and an initiator or activator. The monomers function as a solvent when the adhesive is liquid, but polymerise to form tough, load-bearing solids when activated. The reactive group, which is used to initiate a molecular chain growth in the adhesive, is a polymerisable carbon–carbon double bond. The chain-growth polymerisation reaction proceeds rapidly to harden the adhesive in minutes, and no heat is required during the process. By selection from a wide choice of polymers and monomers, acrylic structural adhesive systems emphasising high shear strength or high peel strength, or both, can be formulated. With acrylics the time to cure an assembly (hand-ling time) can be as short as 30 s or as long as several hours. Typical general-purpose systems will reach handling strength in 15 min and 95% of full strength within 1 h.

The earliest reported durability performance on the present family of new acrylic adhesives was by Zalucha.[277] Steel and aluminium joints were tested for 1000 h in room temperature water, hot condensing humidity, and 5% salt spray with the result that more than 60% strength retention was found for all adherends except SAE 1010 CRS in hot humidity. Minford[278] conducted extensive weathering tests on the first structural acrylics introduced by Hughson Chemical Company in the late 1960s, and a wide range of joint durability responses were elicited. The general durability of these early acrylic bonded joints with minimal surface preparation was generally poor in both water and humidity soaking exposures. However, where the aluminium was acid-etched, the durability response was essentially equal to that of the average commercial two-part epoxy. The shorter curing time for the acrylics, however, was an important advantage. Testing under combination stress and water-soaking conditions was also conducted with the result that the acrylic joints (especially on etched aluminium surfaces) were much superior to two-part epoxy systems.

The present structural acrylic adhesives are recommended by their formulators for bonding to unprepared aluminium surfaces, indicating a significant advance in formulation technology since the first products to be introduced were tested by Minford. The newest product line of Hughson Chemical and some selected so-called 'second generation' acrylics made by licensees of duPont have been evaluated by Minford[121,122,279,280] with the result that the joint strengths were maintained at an acceptable level over two-year soaking periods in liquid water or hot condensing humidity. There was also clear evidence that acrylic-bonded aluminium joints could be considered more durable than two-part epoxy joints in corrosive saltwater exposures. Probably for the basic reasons discussed in Chapters 1 and 3, the most durable joints using structural acrylic adhesives in the testing by Minford[278,279] were achieved with plastic adherends like polyester fibreglass, styrene, ABS, polyvinyl and polycarbonate. As a result, acrylic adhesives can be recommended for bonding dissimilar-material joints between metals and plastics.

Zalucha[281] has also reported the use of acrylic adhesives for plastics and metal bonding in solar energy applications. A complete review of developments in the entire field of acrylic adhesives has been compiled by Martin.[282]

8. CYANOACRYLATES

If the α-carbon of an acrylic monomer has its hydrogen atom substituted by a nitrile (—CN) group, then a highly asymmetric double bond is produced which opens readily in the presence of weakly basic (i.e. electron donor) molecules. The curing is prevented in the adhesive package by the presence of small amounts of an acid inhibitor. When removed from the package, there is enough basicity in water, alcohols or most metal oxides to initiate polymerisation. If an adherend surface is distinctly acid or very dry, the polymerisation may be initiated with a little alcohol. The joint strength ordinarily builds within one minute to the strength expected from a good epoxy. The methyl monomer produces relatively brittle bonds which are obviously subject to cracking failure where slight peel or cleavage stresses are imposed. This adhesive is widely used for small surface assemblies in the electronics area where its speed of cure and sufficient joint strength are distinct advantages.

It has been the general experience of both the formulators and customers involved with using cyanoacrylate adhesives that the majority of bonds will weaken when aged in high humidity atmospheres; therefore, proper bond design to minimise bondline exposure is recommended. Probably the most detailed investigation of cyanoacrylate joint durabilities was conducted by Eastman Kodak Company, who formulated the first practical adhesive (Eastman 910). Although no longer in the business, their pioneering efforts have led to development of the longest exposure time data. Only outdoor weathering conditions have been employed rather than accelerated laboratory humidity soaking, as used for evaluating most other structural adhesives. Also, no durability data for metal-to-metal joints have been provided by Eastman. They did show, however, that outdoor weathering exposures for 4–7 years did not produce a shear strength decrease for dissimilar materials joints like rubber to aluminium or rubber to steel. Phenolic-to-aluminium joints similarly exposed did lose 62% of initial strength within two years.

9. POLYURETHANE ADHESIVES

One of the first known uses of a polyurethane adhesive was for bonding rubber to metal on tank treads in Germany during World War II. While this sounds distinctly like a structural bonding application, more polyurethane adhesives are used for bonding elastomers, fabrics, fibres and thermoplastics to one another than for joining metallic structures. The two-part polyurethane adhesives which have been used to join polyester fibreglass to steel parts in vehicles usually require a primer on the metal for any reasonable durability in weathering performance. Twiss[283] and Smith and Sussman[284] have demonstrated that polyurethane adhesives have superior shear strength and toughness as compared with any other adhesive type at cryogenic temperatures. However, they rapidly lose bond strength above 80–100°C (176–212°F), which limits their use in many structures which operate under natural weathering conditions. For example, a black painted metal surface can exceed this temperature condition due to natural solar exposure alone.

Mobay Corporation[285] has reported development of a very high peel and tensile shear strength polyurethane-type adhesive which has excellent bond moisture resistance even on unprimed steel and aluminium. However, at only 80–100°C (176–212°F) the tensile shear strength has already diminished to half the room temperature value.

Delmonte and Sarna[286] in 1972 reviewed high-strength polyurethane adhesives and reported lap shear values in excess of 34·45 MPa (5000 psi) and peel strengths in excess of 60 lb/in. They proposed that such polyurethanes were more than competitive with the best epoxys. However, etched aluminium joints with these adhesives, when subjected to 95% humidity ageing at only 24°C (75°F), lost as much as 48% of initial joint strength after one month. Delmonte[287] furnished additional information on polyurethanes for metal bonding in 1974.

Recently, a one-part, heat-cured polyurethane was made available by Goodyear that was recommended for bonding steel to polyester fibreglass. It was recommended that a vinyl primer be used on the metal surface. Minford[288] has conducted bond durability tests on this product for joining aluminium to polyester fibreglass and found satisfactory joint durability in water soaking tests to the plastic adherend, but the aluminium interface readily debonded even where the recommended primer was used.

DeLollis[262] has conducted durability testing of polyurethane-joined adherends by both immersion in water and soaking in high humidity. With the concern in mind about the use of any MOCA-curing polyurethanes, Lauer and Boyaner[289] have investigated the formulation and use of non-MOCA (4,4′-methylenebis(2-chloroaniline)) cured polyurethane at −184°C (−300°F).

10. ANAEROBIC ADHESIVES

The unique aspect of these adhesives is that they can be considered as a one-part adhesive which can be kept in the unreacted state for an indefinitely long period as long as oxygen is not excluded from the mix. At the same time there is the potential for curing to form structural strength in any adherend contact situation where oxygen is missing. The existence of certain Redux catalyst systems which are inhibited by oxygen allows mixing with vinyl monomers which can then be polymerised by free-radical initiation where oxygen deprivation (i.e. anaerobic conditions) exists between two adherends. This polymerisation proceeds rather slowly at room temperature, but then is rapidly accelerated under elevated temperature conditions. Cure time also may vary depending on the nature of the adherend since the Redux catalysts are sensitive to the identity of the metal ion present.

The main market for these adhesives has been where the joint once made will be impervious to oxygen migration into the interfacial area. As a result, the locking in place of fasteners has been one major application. Of course, it is possible to exclude oxygen in virtually any joint involving two vapour barrier facings like metal sheeting. These adhesives also sell for prices an order of magnitude greater than epoxys so the assemblies are usually small and the gap clearances narrow.

Pearce[290] has reviewed the successful applications of anaerobics and points out that the durability of anaerobic joints can be questionable unless certain care is exercised. As they are brittle materials there must be concern about the maximum sustained or alternating level of stress which can resist cracking failure at the interface. Certain surfaces may be relatively inactive in supporting the curing process, and primers may be needed to develop proper joint durability. Also, caution is required when trying to joint materials of differing coefficients of thermal expansion, and related to this, sudden thermal shocks which could disrupt the bondline may need to be restricted.

In recent years the Loctite anaerobic products have been modified to offer choices of adhesives which may have better durability for more aggressive weathering or higher temperature resistance. Minford[291] has evaluated two of these new products, Superbonder 324 and Superbonder 325: 324 is considered the more general-purpose product, while 325 is said to have higher temperature resistance above 121–135°C (250–275°F) and superior joint durability in water-soaking environments. Under simultaneous exposure to stress and hot humidity, both adhesives were superior to two-part epoxys but not to the improved structural acrylic adhesives. The general water resistance of these newer anaerobics was much improved over earlier products to the degree that they compared favourably with the typical room temperature-curing epoxys and acrylics. On 3003 aluminium alloy, both anaerobics formed excellent bonds to only vapour-degreased adherends which survived one year in corrosive $3\frac{1}{2}\%$ intermittent salt spray.

A comprehensive review of the technology has been published by Murray.[292]

11. HIGH-TEMPERATURE ADHESIVES

The temperature range and definition for a high-temperature adhesive have generally depended on the needs of the design engineer at the time. The first structural adhesives used in aircraft were satisfactory for those times, but the need for an adhesive to sustain faster aircraft designs and higher metal skin temperatures pushed the range for so-called high-temperature-resistant adhesives still higher. The epoxy–phenolics achieved this goal in the 1950s, but soon the need was for adhesives with longer survival times at temperatures up to 288°C (550°F) which was limiting to the epoxy phenolics.

As the space age dawned, designers began to talk about the need for adhesives which would be suitable for cryogenic 20 K (−424°F) to high temperature 589 K (600°F) service. From the early 1960s an international effort began to develop a new generation of polymers for use above 316°C (600°F). Initially the efforts revolved about fluorocarbon or silicone elastomer improvements, but many entirely new heterocyclic aromatic polymers such as the poly(imide), poly(benzimidazole), and poly(phenylquinoxaline) types began to appear. Some of these definitely showed the ability to retain useful joint strength levels after extended time in air at temperatures in the 316–538°C (600–1000°F) range where the earlier high-modulus polymer adhesives suffered rapid thermal or oxidative degradations. The first development seemed to involve the production of a linear prepolymer which could be dissolved in certain strong solvents. After this solvent is evaporated from the bondline, an internal condensation to form a complicated ring structure is promoted by heating the joint under relatively high pressures to the temperature range 204–316°C (400–600°F). It can readily be demonstrated that a temperature condition which could completely degrade an epoxy–phenolic adhesive within 1–10 h can be survived with good strength retention for 1000 h by these kinds of cured polymers. At the same time the epoxy–phenolic adhesive can maintain a superior level of joint strength following exposure in various water-soaking service conditions, as shown by Twiss.[283]

There are major difficulties associated with these high-temperature-resistant adhesives which have diminished their use, for instance their high cost of manufacture, difficulties associated with handling and curing, and the problems involved in eliminating volatiles during cure to obtain void-free bondlines. The polyimide-type polymers have been the current frontrunners, partly due to more extensive research conducted on these polymers because of their various non-military application potentials.

Durability evaluations of these high-temperature adhesives has generally taken a different direction than for the other types of structural adhesives. Measurements of their properties are predictive of their performance: data on how high a temperature they can survive or how long they can survive at given temperatures are predictive of their durability. Although it is assumed that water will not be present at higher service temperatures, it is certainly logical to assume that periods of inactivity will exist when the bondlines may see ambient conditions. Wrasidlo[293] has measured the thermal properties of polyimides, poly(phenylene ethers) and polyquinoxalines, including their transition and relaxation characteristics. Hergenrother[294] has looked at the poly(phenyl-*as*-triazines) and poly(phenylquinoxalines) for adhesive and composite matrices. Aponyi and Delano[295] have investigated the poly(amidazoquinazolines) for use as matrices in high-performance reinforced composites. Roper[296] demonstrated a very broad temperature range for maintaining high shear strength by BR34/FM34 polyimide. In comparison, Vaughan and Jones[297] showed P4/A5F poly(imide) had significant property and processing advantages over the BR34/FM34 system. Vaughan and Sheppard[298] conducted comparative testing on the above two systems, a Boeing poly(phenylquinoxaline) and duPont NR1508 polyimide. Test procedures included static shear strength and thermal shock from 20 K (−423°F) to 589 K (600°F) and ageing and 50% stressing at 477 K (400°F) to 589 K (600°F). At the latter temperature the P4/A5F joints showed significantly higher strength retention. With stressing and 589 K (600°F) ageing, the poly(phenylquinoxaline) was superior. The static test results at 20 K (−423°F) showed highest strength values for the BR34/FM34 system. It seems that best compromise adhesives must be sought for the particular service conditions envisaged.

Mention was made earlier in this section of the problems of experiencing significant numbers of voids in the bondline using these high-temperature-resistant polymers. St Clair and St Clair[299] recently announced development of LARC-TPI, a linear thermoplastic poly(imide) which can be imidised and freed of volatiles at the unusually low temperature of 230°C (446°F) and then processed as a thermoplastic. These linear poly(imides) are attractive to the aerospace industry because of their unusual toughness with flexibility over a wide temperature range. St Clair and St Clair[144] have also been studying the effects of added elastomers, like amine-terminated acrylonitrile–butadiene copolymers or silicones, on the thermal stability, adhesive strength and fracture toughness of a high-temperature addition poly(imide) adhesive. The rubber elastomers addition sacrificed some high-temperature shear strength, but the high-temperature strengths of the poly(imide) with silicone elastomers actually improved with ageing at 505 K (450°F). The ambient- and elevated-temperature fracture toughness of the poly(imide) was improved three- to five-fold with the rubber additions while the silicone had no particular influence.

Considerable efforts have also been directed towards developing structural foams for use at cryogenic to ultrahigh temperatures. Papers by Allen and Yates,[300] Segal,[301] and Kimmel[302] are representative.

12.   CONCLUDING REMARKS

The importance of establishing the durability potential for any structural adhesive joining system is mandatory for obtaining the acceptance of adhesive bonding as the joining process in a manufacturing situation. While preliminary designs can be predicted on the initial joint strength data offered by the adhesive manufacturer, the ultimate design must take into account how the joint system will respond to its long-term service environment. The outstanding technical progress in recent years towards the goal of obtaining truly durable adhesive bonding with a wide variety of differing material adherends has been triggered by the increasing awareness of the need for such data on the part of the design engineer. In addition, there has been a significant advance in the basic understanding of what forces are acting in the interfacial area and how the weathering process interacts with the adhesive attachment. It is anticipated that as the state-of-the-art continues to advance, the credibility of adhesive bonding as a structural method for joining in manufacturing will continue to improve, resulting in a substantial increase in the use of structural bonding in the future.

REFERENCES

1. Baterip, B. O., *Int. J. Adhesion* (July 1981), 233.
2. Kaelble, D. H., *Physical Chemistry of Adhesion.* Wiley Interscience, New York (1971).
3. DeBruyne, N. A. and Houwink, R. (eds), *Adhesion and Adhesives*, Elsevier, Amsterdam (1951).
4. Wake, W. C., *Adhesion and the Formulation of Adhesives*, Applied Science Publishers, London (1976).
5. Bikerman, J. J., *The Science of Adhesive Joints*, Academic Press, New York (1961).
6. Kinloch, A. J., *J. Mater. Sci.*, **15** (1980), 2141.
7. Huntsberger, J. R., *J. Adhesion*, **1** (1976), 289.
8. Mittal, K. L., *Adhesion Science and Technology*, ed. L. H. Lee, Plenum Press, New York (1975), p. 129.
9. Good, R. J., *Surface Colloid Sci.*, **11** (1979), 1.
10. Wu, S., *Polymer Blends*, Vol. 1, ed. D. R. Paul and S. Newman, Academic Press, New York (1978), Chapter 6.
11. Huntsberger, J. H., in *Treatise of Adhesion and Adhesives*, Vol. 5, ed. R. L. Patrick, Marcel Dekker, New York (1980), Chapter 1.
12. Girifalco, L. A. and Good, R. J., *J. Phys. Chem.*, **61** (1957), 904.
13. Fowkes, G. M., *Ind. Eng. Chem.*, **56**(12) (1964), 40.
14. Gardon, J. L., in *Encyclopedia of Polymer Science and Technology*, Vol. 3, ed. H. F. Mark, Interscience, New York (1965).
15. DeBye, P. J., *Adhesion and Cohesion*, ed. A. A. Weiss, Elsevier, New York (1962).
16. Padday, J. F. and Uffendell, N. D., *J. Phys. Chem.*, **72** (1968), 1407.
17. Good, R. J., *J. Adhesion*, **8** (1976), 1.
18. Bijlmer, P. F. A., *J. Adhesion*, **5** (1973), 319.
19. Evans, J. R. G. and Packham, D. E., *J. Adhesion*, **10** (1979), 177.
20. Eick, J. D., Good, R. J., Newmann, A. W. and Fromer, J. R., *J. Adhesion*, **3** (1971), 23.
21. Claret, P. A., in *Aspects of Adhesion 3*, ed. D. J. Alner, Univ. of London Press (1965), pp. 9–31.
22. McGuire, E. P., *American Adhesive Index*, Padric Publishing, Mountainside, New Jersey (1962).
23. Katz, I., *Adhesive Materials*, revised by C. V. Cagle, Foster Publishing, Long Beach, California (1971).
24. *Adhesives in Modern Manufacturing*, Society of Manufacturing Engineers, Dearborn, Michigan (1970).
25. Cagle, C. V., *Handbook of Adhesive Bonding*, McGraw-Hill, New York (1973).

26. Skeist, I., *Handbook of Adhesives*, 2nd edn, Van Nostrand–Reinhold, New York (1977).
27. Guttmann, W. H., *Concise Guide to Structural Adhesives*, Reinhold Publishing, New York (1961).
28. Cagle, C. V., *Adhesive Bonding Techniques and Applications*, McGraw-Hill, New York (1968).
29. McGuire, E. P., *Adhesive Raw Materials Handbook*, Padric Publishing, Mountainside, New Jersey (1964).
30. *Adhesives*, 1978/1979 Book A, Cordura Publications, San Diego, California.
31. *Adhesives*, 1978/1979 Book B, Cordura Publications, San Diego, California.
32. *Adhesives Red Book*, Palmerton Publishing, Atlanta, Georgia (1982).
33. Bolger, J. C., in *Treatise on Adhesion and Adhesives*, Vol. 3, ed. R. L. Patrick, Marcel Dekker, New York (1973), Chapter 1.
34. Orowan, E., *Rept. Progr. Phys.*, **12** (1949), 185.
35. Schwarzl, F. and Staverman, A. S., *Die Physik der Hochpolymeren*, Vol. IV, ed. H. A. Stuart, Springer, Berlin (1956), pp. 164–213, 226–35.
36. Wolock, I., Kies, J. A. and Newman, S. B., *Fracture*, ed. B. L. Averbach *et al.*, Wiley, New York (1959), pp. 250–64.
37. Bueche, A. M. and Berry, J. P., *Fracture*, ed. B. L. Averbach *et al.*, Wiley, New York (1959), pp. 245–79.
38. Bueche, F., *Rubber Chem. Technol.*, **32** (1959), 1269.
39. Nielsen, L. E., *The Mechanical Properties of Polymers*, Reinhold, New York (1962), pp. 122–37.
40. Stromberg, R. R., in *Treatise on Adhesion and Adhesives*, Vol. 1, ed. R. L. Patrick, Marcel Dekker, New York (1967), Chapter 3.
41. Jenckel, E. and Rumbach, B., *Z. Electrochem.*, **55** (1951), 612.
42. Simha, R., Frisch, H. L. and Eirich, F. R., *J. Phys. Chem.*, **57** (1953), 584.
43. Frisch, H. L., Simha, R. and Eirich, F. R., *J. Phys. Chem.*, **21** (1953), 365.
44. Frisch, H. L. and Simha, R., *J. Phys. Chem.*, **58** (1954), 507.
45. Frisch, H. L. and Simha, R., *J. Phys. Chem.*, **27** (1957), 702.
46. Frisch, H. L., *J. Phys. Chem.*, **59** (1955), 633.
47. Gilliland, E. R. and Guttoff, E. B., *J. Phys. Chem.*, **64** (1960), 407.
48. Silberberg, A. J., *J. Phys. Chem.*, **66** (1962), 1872.
49. Silberberg, A. J., *J. Phys. Chem.*, **66** (1962), 1884.
50. Patat, F. and Schliebener, C., *Makromol. Chem.*, **44–46** (1961), 643.
51. Patat, F. and Estupinen, L., *Makromol. Chem.*, **49** (1961), 182.
52. Patat, F., Killmann, E. and Schliebener, C., *Makromol. Chem.*, **49** (1961), 200.
53. Killmann, E. and Schneider, G., *Makromol. Chem.*, **57** (1962), 212.
54. White, H. W., Godwin, L. M. and Wolfram, J., *J. Adhesion*, **9** (1978), 237.
55. Kaelble, D. H., *Treatise on Adhesion and Adhesives*, Vol. 1, ed. R. L. Patrick, Marcel Dekker, New York (1967), Chapter 6.
56. Huntsberger, J. R., *J. Polym. Sci. A1* (1963), 1339.

57. Bright, W. M., *Adhesion and Adhesives*, Wiley, New York (1954), p. 130.
58. Huntsberger, J. R., *Treatise on Adhesion and Adhesives*, Vol. 1, ed. R. L. Patrick, Marcel Dekker, New York (1967), Chapter 4.
59. Huntsberger, J. R., *J. Polym. Sci.*, A1 (1963), 2241.
60. Kemball, C., *Adhesion and Adhesives*, ed. H. T. Clark, B. Savage and J. E. Rutzler, Wiley, New York (1954), p. 70.
61. Taylor, D., Jr, and Rutzler, J. E., Jr, *Ind. Eng. Chem.*, **50** (1958), 928.
62. Casimer, H. B. G., *Konik, Ned, Akad. Wetenschap.*, *Proc. Ser.*, **B51** (1948), 795.
63. Lifshitz, E. M., *Zh. Eksperim. i. Teor. Fiz.*, **21** (1951), 94.
64. Zisman, W. A., *Adhesion and Cohesion*, ed. P. Weiss, Elsevier, Amsterdam (1962).
65. Zisman, W. A., *Contact Angle, Wettability and Adhesion*, ed. R. F. Gould, American Chemical Society (1964), p. 44.
66. Zisman, W. A., *Ind. Eng. Chem.*, **55** (1963), 18.
67. Zisman, W. A., *Advan. Chem. Ser.*, **43** (1964), 1.
68. Fowkes, F. M., *J. Phys. Chem.*, **67** (1963), 2538.
69. Fowkes, F. M., *Advan. Chem. Ser.*, **43** (1964), 99.
70. Good, R. J. and Girifalco, L. A., *J. Phys. Chem.*, **64** (1960), 561.
71. Good, R. J., *Advan. Chem. Ser.*, **43** (1964), 74.
72. Johnson, R. E., Jr, and Dettre, R. H., *Advan. Chem. Ser.*, **43** (1964), 112.
73. Dettre, R. H. and Johnson, R. E., Jr, *Advan. Chem. Ser.*, **43** (1964), 136.
74. Sharpe, L. H. and Schonhorn, H., *Advan. Chem. Ser.*, **43** (1964), 189.
75. Huntsberger, J. R., *J. Polymer Sci.*, **43** (1960), 581.
76. Eley, D. D. and Tabor, D., *Adhesion*, ed. D. D. Eley, Oxford University Press (1961), p. 9.
77. Baker, F. S., *J. Adhesion*, **10** (1979), 107.
78. Voyutskii, S. S., *Autohesion and Adhesion of High Polymers*, Wiley–Interscience, New York (1963), p. 127.
79. Voyutskii, S. S. and Vakula, V. L., *J. Appl. Polym. Sci.*, **7** (1963), 475.
80. Voyutskii, S. S., Vakula, V. L., Smeloya, N. I. and Tutorskii, I. A., *Vysokomolekul. Soedin.*, **2** (1960), 1671.
81. Flory, P. J., *Principles of Polymer Chemistry*, Cornell Univ. Press, Ithaca, New York (1953), Chapter 13.
82. Hughes, L. J. and Britt, G. E., *J. Appl. Polym. Sci.*, **5** (1961), 337.
83. Dobry, A. and Boyer-Kawenoki, F., *J. Polym. Sci.*, **2** (1947), 90.
84. Derjaguin, B. V., Krotova, N. A., Karassev, D., Kirillova, Y. M. and Alanikova, Y., *Proc. 2nd Intern. Congr. Surface Activity, London* (1957), Vol. III, p. 417.
85. Derjaguin, B. V. and Smilga, V. P., *Proc. 3rd Intern. Congr. Surface Activity, Cologne* (1956), Vol. II Sec. B, p. 349.
86. Skinner, S. M., Savage, R. L. and Rutzler, J. E., Jr, *J. Appl. Phys.*, **24** (1953), 438.
87. Skinner, S. M., *J. Appl. Phys.*, **26** (1955), 498.
88. Skinner, S. M., Gaynor, J. and Sohl, G. W., *WADC Tech. Rept* (1956), pp. 56–158.
89. Zisman, W. A., *Ind. Eng. Chem.*, **55** (1963), 18. ·

90. Kaelble, D. H., *ACS Division of Paint, Plastics, and Printing Ink Chemistry, Cleveland Meeting*, (April 1960), Paper No. 5.
91. Meissner, H. P. and Merrill, E. W., *Am. Soc. Testing Materials Bull.*, **151** (1948), 80.
92. McLaren, A. D. and Siler, C. J., *J. Polym. Sci.*, **4** (1949), 63.
93. Tobolsky, A. J., *Properties and Structure of Polymers*, Wiley, New York (1960).
94. Bueche, F., *Physical Properties of Polymers*, Wiley–Interscience, New York (1962).
95. Good, R. J., *Treatise on Adhesion and Adhesives*, Vol. 1, ed. R. L. Patrick, Marcel Dekker, New York (1967), Chapter 2.
96. Berry, J. P. and Bueche, A. M., in *Adhesion and Cohesion*, ed. P. Weiss, Elsevier, Amsterdam (1962).
97. Bikerman, J. J., in *Recent Advances in Adhesion*, ed. L. H. Lee, Gordon and Breach, New York (1973).
98. Good, R. J., in *Recent Advances in Adhesion*, ed. L. H. Lee, Gordon and Breach, New York (1973).
99. Sharpe, L. H., *J. Adhesion*, **4** (1972), 51.
100. Minford, J. D., *SAMPE Q.*, (July 1978), 18.
101. Sterman, S. and Toogood, J. B., *Adhesives Age*, (July 1965).
102. Laird, J. A., Glass surface chemistry for glass reinforced plastics, *Final Report, Navy Contract W-0679-C (FBM)* (June 1963).
103. Gledhill, R. A. and Kinloch, A. J., *J. Adhesion*, **6** (1974), 315.
104. Barlow, D. A., *The Effect of Dry and Wet Exposure Tests at Elevated Temperatures on the Static Strength of Redux Joints*, Report B-BR-73-4B-27, Alcan International Ltd, Banbury, (1948).
105. Butt, R. I. and Cotter, J. L., *J. Adhesion*, **8** (1976), 11.
106. Mazor, A., Broutman, L. J. and Eckstein, B. H., *Nat. Tech. Conf. SPE* (Preprint) (1976), 77.
107. Farrar, N. R. and Ashbee, K. H. G., *J. Phys. D*, **11** (1978), 1009.
108. Nicholas, J. and Ashbee, K. H. G., *J. Phys. D*, **11** (1978), 1015.
109. Comyn, J., *Developments in Adhesives—2*, ed. A. J. Kinloch, Applied Science Publishers, London (1981), Chapter 8.
110. Minford, J. D., *Metals Eng. Quart.* (November 1972).
111. Minford, J. D., *Treatise on Adhesion and Adhesives*, Vol. 3, ed. R. L. Patrick, Marcel Dekker, New York (1973), Chapter 2.
112. Minford, J. D., *Adhesives Age*, (July 1974), 24.
113. Minford, J. D., *J. Appl. Polym. Symp.*, **32** (1977), 91.
114. Krieger, R. B., Jr, *Symposium on Processing for Adhesive Bonded Structures*, (August 1972), 644.
115. Riel, F. J., *SAMPE J.*, **7** (1971), 16.
116. Rogers, N. Corrosion of adhesive bonded clad aluminum. *S.A.E. National Business Aircraft Meeting, Wichita, Kansas* (March 1972).
117. Dahringer, D. W., *Symposium on Processing for Adhesive Bonded Structures*, (August 1972), 575.
118. Minford, J. D., *Evaluation of Compatibility of Electrically Conductive Epoxies with Aluminum Adherends in Water Exposure Conditions*, unpublished report, Alcoa Laboratories.
119. Ojalvo, I. U. and Eidenoff, H. L., Bond thickness effects upon stresses in

single lap adhesive joints, *S.A.E. Paper No.* 770090, (28 Feb.–4 March 1977).

120. Minford, J. D. and Vader, E. M., Adhesive bonding of aluminum automotive body sheet, *S.A.E. Int. Auto. Eng. Cong. Paper No.* 740078, (February 1974).

121. Minford, J. D., Weldbonding aluminum in the presence of forming lubricants, *S.A.E. Paper No.* 810816, *Passenger Car Meeting, Dearborn, Michigan* (8–12 June 1981).

122. Minford, J. D., Comparative effect of surface contamination on the strength and performance of aluminum spot-welded or adhesive-bonded joints, *ASM/ADDRG Conference—Technological Impact of Surfaces, Relationship to Forming, Welding and Painting, Dearborn, Michigan,* (April 1981).

123. Danforth, M. A. and Sunderland, R. J., *J. Appl. Polym. Sci., Appl. Polym. Symp.,* **32** (1977), 201.

124. DeBruyne, N. A., The extent of contact between glue and adherend, *Bulletin* 168, Tech. Service Dept., Aero Research Ltd, Dufford, Cambridge, (December 1956).

125. Bascom, W. D. The origin and removal of microvoids in filament wound composites, *NRL Report* 6268, (24 May 1965).

126. Plueddemann, E. P., Adhesion through silane coupling agents, *25th SPI Reinforced Plastics Composites Div. Conf., Washington, DC,* (February 1970).

127. Minford, J. D., unpublished report, Alcoa Laboratories.

128. Minford, J. D., unpublished report, Alcoa Laboratories.

129. Sell, W. D., Some analytical techniques for durability testing of structural adhesives, *19th Nat. SAMPE Conf.,* (April 1974).

130. Bodnar, M. J. and Wegman, R. F., *SAMPE J.,* (Aug/Sept. 1969), 51.

131. Minford, J. D., unpublished report, Alcoa Laboratories.

132. Spathis, G. D., Sideridis, E. P. and Theocaris, P. S., *Int. J. Adhesion Adhesives,* (April 1981), 195.

133. Wolfe, H. F., Rupert, C. L. and Schwartz, H. S., Evaluation of several bonding parameters on the random fatigue life of adhesively bonded aluminum joints, *AFWAL-TR-81-3096,* (Aug. 1981).

134. Wake, W. C., *Trans. IRI,* **35**(5) (1959), 146.

135. Wake, W. C., *Adhesion,* ed. D. D. Eley, Oxford University Press, Oxford (1961), p. 202.

136. Bryant, R. W. and Dukes, W. A., *Brit. J. Appl. Phys.,* **16** (1965), 101.

137. Bryant, R. W. and Dukes, W. A., *Applied Polym. Symposia,* Vol. 3, Interscience, New York (1966), p. 81.

138. Bryant, R. W. and Dukes, W. A., *S.A.E. Aeronautic and Space Eng. and Mfg Meeting, Los Angeles,* Ref. 670855 (1967).

139. Foulkes, H., Shields, J. and Wake, W. C., *J. Adhesion,* **2** (1970), 254.

140. Black, J. M. and Blomquist, R. B., *Ind. Eng. Chem.,* **50** (June 1958), 918.

141. Black, J. M. and Blomquist, R. B., *Development of Metal Bonding Adhesives with Improved Heat Resistance Properties,* NACARM 52F19, Forest Products Laboratory, (1952).

142. Black, J. M. and Blomquist, R. B., *Development of Metal Bonding*

dhesives with Improved Heat-resistant Properties, NACARMS 54D01, ...orest Products Laboratory; *Modern Plastics*, **32** (December 1954), 139.

143. Black, J. M. and Blomquist, R. B., *Modern Plastics*, (June 1956).

144. St Clair, A. K. and St Clair, T. L., *Int. J. Adhesion Adhesives*, (July 1981), 249.

145. Griffith, A. A., *Phil. Trans.*, Roy. Soc., **221** (1920), 163.

146. Irwin, G. R., in: *Structural Mechanics*, ed. J. N. Goodier and N. J. Hoff, Pergamon Press, New York (1960).

147. Ripling, E. J., Mostovoy, S. and Patrick, R. L., Application of fracture mechanics to adhesive joints, *Adhesion*, ASTM (1964).

148. Mostovoy, S., Crosley, P. P. and Ripling, E. J., *J. Mater.*, **2**(3) (1967).

149. Mostovoy, S. and Ripling, E. J., Fracturing characteristics of adhesive joints, *Final Report, Contract N00019-75-C-0271*, Materials Research Lab. (1975–76).

150. Ripling, E. J., Mostovoy, S. and Corten, H. T., *J. Adhesion*, **3** (1971), 107.

151. Mostovoy, S. and Ripling, E. J., Fracturing characteristics of adhesive joints, *Final Report, Contract N00019-17-C-0329*, Materials Research Lab. (1972).

152. Mostovoy, S. and Ripling, E. J., *J. Appl. Polym. Sci.*, **15** (1971), 661.

153. Mostovoy, S. and Ripling, E. J., *J. Appl. Polym. Sci.*, **13** (1969), 1083.

154. Mostovoy, S. and Ripling, E. J., *J. Appl. Polym. Sci.*, **15** (1971), 641.

155. Mostovoy, S., Smith, H. R., Lingwall, R. G. and Ripling, E. J., *Eng. Fracture Mech.*, **3** (1971), 291.

156. Ripling, E. J., Mostovoy, S. and Bersh, C. F., *J. Adhesion*, **3** (1971), 145.

157. Mostovoy, S. and Ripling, E. J., *J. Appl. Polym. Symp.*, **19** (1972), 395.

158. Bethune, A. W., Durability of bonded aluminum structure, *19th Nat. SAMPE Symp., Los Angeles*, (April 1974).

159. Marceau, J. A. and Moji, Y., Development of environmentally stable aluminum adhesive bonds, *Document D6-41145*, Boeing Commercial Airplane Co.

160. Common bonding requirements for structural adhesives, *Process Document BAC 5514*, Boeing Commercial Airplane Company.

161. Marceau, J. A. and Moji, Y., Application of fracture mechanics testing to process control for adhesive bonding, *Document G-41141*, Boeing Company, (May 1973).

162. Mostovoy, S. and Ripling, E. J., Fracturing characteristics of adhesive joints, *Final Report, Contract N00019-74-C-0274*, Materials Research Laboratory (January 1974–January 1975).

163. Bascom, W. D., Timmons, D. O. and Jones, R. L., *J. Mater. Sci.*, **10** (1975), 1037.

164. Kinloch, A. J. and Shaw, S. J., *Developments in Adhesives—2*, ed. A. J. Kinloch, Applied Science Publishers, London (1981), Chapter 4.

165. Jennings, C. W., *Recent Advances in Adhesion*, ed. L. H. Lee, Gordon and Breach, New York (1973), p. 469.

166. Mulville, D. R. and Vaishnav, R. N., *Army Symposium on Solid Mechanics*, Army Materials and Mechanics Research Center, Watertown, Maine (1974).

167. Wilcox, R. C. and Jemian, W. A., *Polym. Eng. Sci.*, **13** (1973), 40.
168. Jones, W. B., Kaelble, D. H. and Knauss, W. G., Investigations of fatigue and crack propagation behavior of adhesives, *Air Force Materials Laboratory Report AFML-TR-72-218*, Part I, (October 1972).
169. Jones, W. B. and Kaelble, D. H., Investigation of fatigue and crack propagation behavior of adhesives, *Air Force Materials Laboratory Report AFML-TR-72-218*, Part II, (December 1973).
170. Bascom, W. D., Cottington, R. L. and Timmons, C. O., *Nav. Eng. J.*, (August 1978), 73.
171. Loew, M. H., Fitzgerald, J. M. and Mucciardi, A. N., *Air Force Materials Lab. Final Report AFML-TR-78-206* (December 1978).
172. Clark, H. T., Definition and non-destructive detection of critical adhesive bond-line flaws, *Air Force Materials Lab. Final Report* (July 1978).
173. Cuthrell, R. E., *J. Appl. Polym. Sci.*, **12** (1968), 1263.
174. Zhurkov, S. N., Kuksenko, V. S. and Slutsker, A. I., *Fracture*, Chapman and Hall, London (1969), p. 531.
175. Schliekelmann, R. J., Non-destructive testing of bonded joints—Recent developments in testing systems, *Non-Destructive Testing*, **8** (April 1975), 100. (A/Fokker-VFW, Schiphal-Oost, The Netherlands.)
176. Matzkanin, G. A., Gardner, C. G. and Burkhart, G. I., Improved inspection for bonded C-5A structure, *Southwest Research Institute Report No. 15–1311*, (August 1974).
177. Dukes, W. A. and Kinloch, A. J., Non-destructive testing of bonded joints, an adhesion science viewpoint, *Non-destructive Testing Res. Practice* (1974).
178. Arnold, J. S. and Vincent, C. T., Development of non-destructive tests for structural adhesive bonds, *Standard Research Institute Report*, (December 1959).
179. Rose, J. L. and Meyer, P. A., *Mater. Eval.*, **31**(6) (1973), 109.
180. Rose, J. L., *J. Appl. Polym. Sci., Appl. Polym. Symp.*, **32** (1977), 389.
181. Tattersall, H. G., *J. Appl. Phys.* (1973), 6.
182. Dance, W. E. and Peterson, D. H., *J. Appl. Polym. Sci., Appl. Polym. Symp.*, **32** (1977), 399.
183. McAbee, E. and Levi, D. W., *J. Appl. Polym. Sci.*, **11** (1967), 2067.
184. McAbee, E. and Levi, D. W., Use of a reaction rate method to predict failure times of adhesive bonds at constant stress, *Technical Report 4105*, Materials Engineering Laboratory, Picatinny Arsenal, (December 1970).
185. Frazier, T. B. and Lajoie, A. D., Durability of adhesive bonded joints, *Technical Report AFML-TR-74-26*, Bell Helicopter Company (March 1974).
186. Mulville, D. R. and Vaishnav, R. N., *J. Adhesion*, **7** (1975), 215.
187. Grimes, G. C., Storage time effects on adhesive mechanical properties, *Symposium for Processing Adhesive Bonded Structures, Stevens Inst. of Tech.*, (August 1972).
188. Marceau, J. A. and Scardino, W., Durability of adhesive bonded joints, *TR-75-3*, Air Force Materials Lab., Dayton, Ohio, (Feb. 1975).
189. McMillan, J. C., *Developments in Adhesives—2*, ed. A. J. Kinloch, Applied Science Publishers, London (1981), Chapter 7.

190. Carter, C. F., Durability of adhesive joints, *ASTM, STP No. 401*, (1966), p. 28.

191. Locke, M. C., unpublished data, Boeing Commercial Airplane Company (1980).

192. Eickner, H. W., Environmental exposure of adhesive bonded metal lap joints, *WADC Tech. Report 59–567*, Part 1, (Feb. 1960).

193. Eickner, H. W., Environmental exposure of adhesive bonded metal lap joints, *WADC Tech. Report 59-564*, Part 1, Forest Products Laboratory (1962).

194. Eickner, H. W., Olsen, W. Z. and Lullig, R. M., Resistance of adhesive-bonded metal lap joints to environmental exposure, *Technical Document WADC TR-59-654*, Part II, Forest Products Laboratory, (October 1962).

195. Sharpe, L. H., Some aspects of the permanence of adhesive joints, *Symp. Struct. Adhesives Bonding, Stevens Institute of Technology, Hoboken, NJ*, (September 1965).

196. Carter, G. F., *Adhesives Age*, **10** (October 1967), 32.

197. Yurek, D. A., private communication, 3M Adhesive Coatings and Sealers Division, Test Method C-294, (June 1972).

198. Minford, J. D., Aluminum joint durability in stress and condensing humidity exposure with varying surface pretreatment, *ASM/SAMPE Conf. Specialized Cleaning, Finishing, and Coating Processes, Los Angeles*, (Feb. 1980).

199. Minford, J. D., *Adhesives Age*, (October 1980), 36.

200. Bascom, W. D., *Adhesives Age*, (April 1979), 28.

201. Kreiger, R. B., Jr, Evaluating structural adhesives under sustained load in hostile environments, *5th National SAMPE Technical Conference*, (October 1973).

202. Marceau, J. A., Mojii, Y. and McMillan, J. C., A wedge test for evaluating adhesive bonded surface durability, *SAMPE Symposium* (April 1976).

203. Shannon, R. W. and Thrall, E. W., *1976 Preprints, Symposium on Durability of Adhesive Bonded Structures, US Army Armament R & D Center, Dover, New Jersey*, (October 1976), p. 195.

204. Scardino, W. M. and Marceau, J. A., *Symposium on Durability of Adhesive Bonded Structures, US Army Armament R & D Center, Dover, New Jersey*, (October 1976), p. 81.

205. Cherry, B. W. and Thomson, K. W., *Fracture mechanics and technology*, ed. G. C. Sih and C. L. Chow, Vol. 1, Sijthoff and Noordhoff, Amsterdam (1977), p. 723.

206. Bascom, W. D., Gadomski, S. T., Henderson, C. M. and Jones, R. L., *J. Adhesion*, **8** (1977), 213.

207. Coble, R. L. and Parikh, N. M., *Fracture, an Advanced Treatise*, ed. H. Liebowitz, Vol. 7, Academic Press, New York (1972), p. 243.

208. Patrick, R. L., Braun, J. A., Cameron, N. M. and Gehman, W. G., *Appl. Polym. Symp.*, **16** (1971), 87.

209. Chen, J. M., Sun, T. S., Venables, J. D. and Hopping, R., *22nd National SAMPE Symposium*, (April 1977), p. 25.

210. Bascom, W. D., *J. Adhesion*, **2** (1970), 161.

211. Falconer, D. J., Walker, P. and MacDonald, N. C., *Chem. Ind.* (1964), 1230.
212. DeLollis, N. J., *Appl. Polym. Sci.* (June 1967).
213. Kerr, C., MacDonald, N. C. and Orman, S., *J. Appl. Chem.*, **17** (1967), 62.
214. Scott, J. A., Durability of bonded joints, *ASTM STP 401* (1966), 16.
215. Orman, S. and Kerr, C., *Aspects of Adhesion*, ed. D. J. Alner, University of London Press, London (1971), p. 64.
216. Brockmann, W., The environmental resistance of metal bonds in new industries and applications for advanced materials technology, *19th Nat. SAMPE Symp. Exhibition, Azuza, California* (1974).
217. Brockmann, W., *Bicent. Mater. Progr.—SAMPE* (1976), 383.
218. Kinloch, A. J., Interfacial fracture mechanical aspects of adhesive bonded joints, *AGARD Lecture Series 102*, Wright–Patterson Air Force Base, Dayton, Ohio, (October 1979).
219. Gledhill, R. A., Kinloch, A. J. and Shaw, S. J., *J. Adhesion*, **11** (1980), 3.
220. Solomon, G., *Adhesion and Adhesives*, Vol. 1, ed. R. Houwink and G. Solomon, Elsevier, New York (1965), p. 325.
221. Minford, J. D., *Evaluation of Long-term Bond Strength of Aluminum Joints Fabricated with Aircraft Type Structural Bonding Adhesive Tape and Film*, unpublished report, Alcoa Laboratories, (5 April 1972).
222. Catchpole, E. J., Some recent European developments in the structural adhesives field, *Symposium on Structural Adhesive Bonding*, Vol. 1, Stevens Inst. of Tech. (Sept. 1965), p. 331.
223. Koetsier, J., *Proc. Nat. SAMPE Technical Conf.*, **7** (1975), 126.
224. DeLollis, N. J., Durability of structural adhesives (a review), *22nd National SAMPE Symp. Exhibition, San Diego*, (April 1977), 673.
225. Olson, W. Z. *et al.*, Resistance of adhesive bonded metal lap joints to environmental exposure, *Report No. WADC-TR-59-564*, (October 1962).
226. Hockney, M. D., *Technical Reports* (a) 70081 (1970); (b) 72100 (1972); (c) 73013 (1973), Royal Aircraft Establishment, Farnborough.
227. Rogers, N. L., *J. Appl. Polym. Sci., Appl. Polym. Symp.*, **32** (1977), 37.
228. Eickner, A. W. and Schowalter, W. E., A study of methods for preparing clad 24S-T3 aluminum alloy sheet surfaces for adhesive bonding, Part I, *USAF-PO-(33-038) 49-4696E Report No. 1813*, (May 1950).
229. Eickner, A. W. and Schowalter, W. E., Studies of some of the more promising cleaning methods in treatment of contaminated surfaces of clad 24S-T3 aluminum alloy sheet, Part II, *USAF-PO-(33-038) 49-4696E Report No. 1813*, (May 1950).
230. Novelli, D. C., L-1011 adhesive bonding system, *Symposium on Adhesive and Bonding Processes for the Automobile and Aircraft Industries, ASTM D-14 Sponsored* (October 1979).
231. Bolger, J. C. and Michaels, A. S., *Interface Conversion for Polymer Coatings*, ed. P. Weiss, Elsevier, New York (1969), Chapter 1.
232. Smarnook, W. H. and Bonotto, S., *Polym. Sci. Eng.*, **8**(1) (1968), 41.
233. Brown, H. P. and Anderson, J. F., *Handbook of Adhesives*, ed. I. Skeist, Reinhold, New York (1962), Chapter 19.
234. Bethune, A. W., *SAMPE J.*, **11**(3), (July/Aug./Sept. 1975).

235. Wolfe, H. F., Rupert, C. L. and Schwartz, H. S., Evaluation of several bonding parameters on the random bending fatigue life of adhesively bonded aluminum joints, *AFWAL-TR-81-3096*, (August 1981).
236. Minford, J. D., *Evaluation of Epon 422 with Aluminum Adherends*, unpublished report, Alcoa Laboratories (1969).
237. Kuno, J. K., Comparison of adhesive classes for structural bonding of ultrahigh and cryogenic temperature extremes, *7th National SAMPE Symp. Proc.*, (May 1964).
238. Lewis, A. F. and Saxon, R., *Epoxy Resins*, ed. H. Kakwichi, Marcel Dekker, New York (1969), Chapter 10.
239. Lee, H. and Neville, K., *Handbook of Epoxy Resins*, McGraw-Hill, New York (1967).
240. Bruins, P. F., *Epoxy Resin Technology*, Interscience, New York (1968).
241. Eickner, H. W., Weathering of adhesive bonded lap joints of clad aluminum alloys, *Technical Report 54-447*, Part 1, Forest Products Laboratory, WADC (1955).
242. Eickner, H. W., Weathering of adhesive bonded lap joints of clad aluminum alloys, *Technical Report 54-447*, Part 2, Forest Products Laboratory, WADC (1957).
243. Eickner, H. W., Weathering of adhesive bonded lap joints of clad aluminum alloys, *Technical Report 54-447*, Part 3, Forest Products laboratory, WADC (1958).
244. Minford, J. D., *Int. J. Adhesion Adhesives* (January 1982), 25.
245. Bodnar, M. J. and Wegman, R. F., Effects of outdoor aging on unstressed adhesive bonded aluminum-to-aluminum lap shear joints, *Technical Report No. 3689*, Picatinny Arsenal, Dover, Delaware, USA (May 1968).
246. Cotter, J. L., *Developments in Adhesives—1*, ed. W. C. Wake, Applied Science Publishers, London (1977), Chapter 1.
247. Schlies, A. P., Long term aging of epoxy bonds, *Report ACF-412-329*, ACF Industries (June 1967).
248. Levi, D. W., *et al.*, *SAMPE Q.*, (April 1976), 1.
249. Levi, D. W., *J. Appl. Polym. Sci., Appl. Polym. Symp.*, **32** (1977), 189.
250. Tanner, W. C., *Symposium on Structural Adhesives Bonding, Stevens Inst. of Tech.* (Sept. 1965), Vol. 1, pp. 1–45.
251. Tanner, W. C., Manufacturing processes with adhesive bonding, *Symposium for Processing Adhesive Bonded Structures, Stevens Inst. of Tech.* (August 1972).
252. Wangsness, D. A., *J. Appl. Polym. Sci., Appl. Polym. Symp.*, **32** (1977), 291.
253. Rowe, E. H., Siebert, A. R. and Drake, R. S., *Mod. Plastics*, **47** (1970), 110.
254. Siebert, A. R. and Riew, C. K., *ACS Preprints, Organic Coatings and Plastics Div.*, **31** (1971), 555.
255. Sultan, J. N., Laible, R. C. and McGarry, F. J., *J. Appl. Polym. Sci.*, **6** (1971), 127.
256. Sultan, J. N. and McGarry, F. J., *Polym. Eng. Sci.*, **13** (1973), 29.
257. Riew, C. K., Rowe, E. H. and Siebert, A. R., Rubber toughened thermosets, *Symposium on Toughness and Brittleness of Plastics, Div. Org.*

Coatings and Plastics Chem., *168th ACS Meeting, Atlantic City*, (Sept. 1974).

258. Bascom, W. D., Cottingham, R. L., Jones, R. L. and Peyser, P., *J. Appl. Polym. Sci.*, **19** (1975), 2545.

259. Hughes, E. J. and Rutherford, J. L., *J. Appl. Polym. Sci.*, *Appl. Polym. Symp.*, **32** (1977), 353.

260. Grimes, G. C., *J. Appl. Polym. Sci.*, *Appl. Polym. Symp.*, **32** (1977), 247.

261. Schwartz, H. S., *J. Appl. Polym. Sci.*, *Appl. Polym. Symp.*, **32** (1977), 65.

262. DeLollis, N. J., Properties of silicones, polyurethanes, and structural epoxy adhesives, *Sandia Laboratories Report SLA-73-0365* (October 1973).

263. DeLollis, N., *Adhesives Age*, (Jan. 1969), 25.

264. Minford, J. D., *Treatise on Adhesion and Adhesives*, Vol. 5, ed. R. L. Patrick, Marcel Dekker, New York (1981), Chapter 3.

265. Peterka, J., *Adhäsion*, **7/8** (1975), 200.

266. Askins, D. R. and Schwartz, H. S., *9th National SAMPE Technical Conference, Atlanta, Georgia* (October 1977), Vol. 9, p. 329.

267. Askins, D. R. and Schwartz, H. S., *J. Appl. Polym. Sci.*, *Appl. Polym. Symp.*, **32** (1977), 217.

268. Shannon, R. W. and Thrall, E. W., Jr, *J. Appl. Polym. Sci.*, *Appl. Polym. Symp.*, **32** (1977), 131.

269. McMillan, J. C., Bonded joints and preparation for bonding, NATO, *AGARD Lecture Series* 102, (March 1979), Paper 7.

270. Wegman, R. F., *J. Appl. Polym. Sci.*, *Appl. Polym. Symp.*, **32** (1977), 1.

271. Scardino, W. M. and Marceau, J. A., *J. Appl. Polym. Sci.*, *Appl. Polym. Symp.*, **32** (1977), 51.

272. Reinhardt, T. J., Jr, *Adhesion 2*, ed. K. W. Allen, Applied Science Publishers, London (1978), Chapter 6.

273. Noland, J. S., *Adhesion Science and Technology*, Part A, ed. L. Lieng-Huang, Plenum Press, New York (1975), p. 413.

274. Falcone, A. S. and Miller, J. E., *J. Appl. Polym. Sci.*, *Appl. Polym. Symp.*, **32** (1977), 247.

275. Spencer, R. W., Adhesive primers for multiple stage bonding and surface corrosion protection, *Symposium for Processing Adhesive Bonded Structures, Stevens Inst. of Tech.*, (August 1972).

276. Aker, S. C., The function of adhesive primers in aircraft bonding of aircraft structures, *Symposium for Processing Adhesive Bonded Structures, Stevens Inst. of Tech.* (August 1972).

277. Zalucha, D. J., *Adhesives Age* (February 1972), 21.

278. Minford, J. D., *Summary Report on the Durability of Aluminum Joints Fabricated with a Variety of Commercial Room Temperature Curing Rapid Set Acrylic Adhesives*, unpublished report, Alcoa Laboratories, (Aug. 1974).

279. Minford, J. D., Adhesive joining aluminum to engineering plastics. I, Polyester fiberglass composite, *International Conference on Physicochemical Aspects of Polymer Surfaces, New York City* (Aug. 1981).

280. Minford, J. D., Adhesive joining aluminum to engineering plastics. II, Engineering grade styrene and cross-linked styrene, *International Confer-*

ence on *Physicochemical Aspects of Polymer Surfaces, New York City,* (Aug. 1981).

281. Zalucha, D. J., *Plastics Design and Processing,* (May 1981), 25.
282. Martin, F. R., *Developments in Adhesives—1,* ed. W. C. Wake, Applied Science Publishers, London (1977), Chapter 6.
283. Twiss, S. B., *Structural Adhesives Bonding,* ed. H. J. Bodner, Interscience, New York (1966), p. 455.
284. Smith, M. B. and Sussman, S. E., Development of adhesives for very low temperature application, *Summary Report, Narmco Research and Development (NASA Contract NAS-8-1565)* (May 1963).
285. *A Mondur: Multron Adhesive Formulation,* Chemical Bulletin, Mobay Chemical Corp., Pittsburgh.
286. Delmonte, J. and Sarna, E. C., *Symposium on Processing for Adhesive Bonded Structures, Stevens Inst. of Tech., Hoboken, NJ,* (August 1972), 320.
287. Delmonte, J., *19th Nat. SAMPE Symp. Exhibition,* Vol. 19 (1974), p. 1.
288. Minford, J. D., *Evaluation of One-part Heat Curing Polyurethane for Joining Aluminum to Polyester Fiberglass,* unpublished report, Alcoa Laboratories (1982).
289. Lauer, H. K. and Boyaner, M. R., *J. Appl. Polym. Sci., Appl. Polym. Symp.,* **32** (1977), 301.
290. Pearce, M. B., *Symposium on Processing for Adhesive Bonded Structures, Stevens Inst. of Tech., Hoboken, NJ,* (August 1972), p. 329.
291. Minford, J. D., *Evaluation of New Commercial Adhesive Products: Part I, Superbonder Products of Loctite,* unpublished report, Alcoa Laboratories, (January 1980).
292. Murray, B. D., *J. Appl. Polym. Sci., Appl. Polym. Symp.,* **32** (1977), 411.
293. Wrasidlo, W., *19th Nat. SAMPE Symp. Exhibition,* Vol. 19 (1974), p. 120.
294. Hergenrother, P. M., *19th Nat. SAMPE Symp. Exhibition,* Vol. 19 (1974), p. 146.
295. Aponyi, T. J. and Delano, C. B., *19th Nat. SAMPE Symp. Exhibition,* Vol. 19 (1974), p. 178.
296. Roper, W. D., Spacecraft adhesives for long life and extreme environments, *NASA TR32-1537* (1971).
297. Vaughan, R. W. and Jones, R. J., The development of autoclave processsable thermally stable adhesives for titanium alloy and graphite composite structures, *NASA CR-112003* (December 1971).
298. Vaughan, R. W. and Sheppard, C. H., Cryogenic temperature structural adhesives, *19th Nat. SAMPE Symp. Exhibition,* Vol. 19 (1974), p. 7.
299. St Clair, A. K. and St Clair, T. L., *SAMPE Q.,* (October 1981), 20.
300. Allen, J. D. and Yates, C. I., Preparation and characterization of high temperature syntactic foams, *19th Nat. SAMPE Symp. Exhibition,* Vol. 19 (1974), p. 42.
301. Segal, C. L., *19th Nat. SAMPE Symp. Exhibition,* Vol. 19 (1974), p. 51.
302. Kimmel, B. G., *19th Nat. SAMPE Symp. Exhibition,* Vol. 19 (1974), p. 55.
303. Kausen, R. C., High and low temperature adhesives—where do we stand? *Materials Symposium, Nat. SAMPE 7th, Los Angeles* (1964).

5

Aluminium Adherends

D. M. Brewis

Leicester Polytechnic, Leicester, UK

ABBREVIATIONS

AES	Auger electron spectroscopy
DCB	double cantilever beam
DGEBA	diglycidyl ether of bisphenol A
ISS	ion scattering spectroscopy
R.H.	relative humidity
SEM	scanning electron microscopy
SIMS	secondary ion mass spectroscopy
TEM	transmission electron microscopy
XPS	X-ray photoelectron spectroscopy

1. INTRODUCTION

The adhesive bonding of aluminium alloy components in aircraft has been widely used for many years, as discussed in Chapter 8. Initially bonding was limited mainly to secondary structures but more recently applications involving primary structures have been made.[1] The increasing interest in adhesive bonding in aircraft has been demonstrated by the construction and testing of an extensively bonded fuselage by the McDonnell Douglas Corporation under the Primary Adhesively Bonded Structures Technology (PABST) programme. The advantages of adhesives in aircraft construction include weight saving, improved aerodynamics and reduced cost of manufacture.[2] Although aerospace applications have been the most important example for bonding aluminium parts, an increasing use of aluminium in road vehicles is likely to provide another large application for adhesive bonding.

It is relatively easy to obtain high initial joint strengths with aluminium but it is clearly vital that most of the joint performance is retained for the life of the structure. Lack of confidence in long-term durability has been the main limiting factor in the use of adhesives in aircraft construction. McMillan[3] notes that this lack of confidence is because of some disappointing applications of aluminium structural bonding where environmentally induced disbonds and associated corrosion resulted in major maintenance and repair problems. To achieve the necessary confidence in bonding it is necessary to show that a structure will retain most of its initial properties after being exposed to appropriate environments.

To ensure the safety of a bonded structure, extensive durability testing must be carried out and the tests should relate to in-service performance. Various tests that have been used to assess durability are described in the next section. The life of a bonded structure, e.g. an aircraft, may be about 30 years and therefore accelerated tests must be used if new adhesives are to be adopted in a reasonable period. However, as with all accelerated tests, great caution must be exercised in the interpretation of the results.

Aluminium joint strengths may deteriorate for a variety of reasons but the presence of water can be especially detrimental (see Chapters 1 and 3). Other liquids that may cause problems include hydrocarbon fuels, brake fluids and antifreeze. Ions may also cause serious durability problems. The effect of salt on bonded structures in service is an obvious example but ions originating from various sources prior to bonding can also have a serious effect (see sections 3.1 and 4.6, for example). Elevated temperature and applied stress may markedly reduce durability, especially in the presence of water.

The durability of aluminium joints is also much affected by factors which are under the control of the manufacturer and his customer. These variables, which include the alloy composition, the pretreatment used, the primer and the adhesive, are examined in section 4.

There is therefore a complex range of factors affecting the durability of aluminium joints. The object of this chapter is to attempt to clarify the relative importance of the factors and to examine their interrelationships when bonding aluminium adherends. In addition the possible mechanisms by which loss in joint strength can occur are discussed.

2. ASSESSMENT OF DURABILITY

2.1. General

The durability of adhesive joints may be assessed either by means of laboratory tests or by outdoor trials. The majority of these procedures have been directed towards aerospace applications,[3] but such testing will become increasingly important in other industries as the adhesive bonding of aluminium becomes more widely used. It is clearly essential that the tests employed should correlate with in-service conditions and therefore great care must be taken in the selection of joint type, the environment and the joint loading procedure.

2.2. Laboratory Tests

2.2.1. Development of Testing Procedures

In laboratory tests, many different types of adhesive joint have been used to assess durability. Until relatively recently, single lap, double lap or peel tests were usually used (see Chapter 1, Fig. 1), the joints having first been subjected to a definite environment, e.g. water vapour at 100% R.H. at 50°C for 1000 h. The results of such tests are often expressed as a percentage retention of the initial joint strength.

In recent years the use of self-stressing test methods, i.e. those not requiring the application of an external load, have become important. Two important examples of such tests are given in Fig. 1, namely those using thick and thin double-cantilever beam specimens. The latter is frequently termed the Boeing wedge test and has the advantage that it is inexpensive. As discussed in Chapter 1, the joint is stressed by driving a wedge into the bondline, the rate of crack propagation is measured and the mode of failure, i.e. whether interfacial or cohesive, is assessed.

A traditional assessment of durability is given in the United States Federal Specification MMM-A-132 which describes several environmental tests. These include determining the residual lap shear strengths of aluminium joints after exposure to various conditions, e.g. 30 days exposure to $9.5 \pm 1.7°C$ and 95–100% R.H., 30 days exposure to salt spray at room temperature, immersion in distilled water for 30 days, or immersion in jet fuel, antifreeze or hydraulic oil for seven days.

A satisfactory performance was based upon the residual strengths

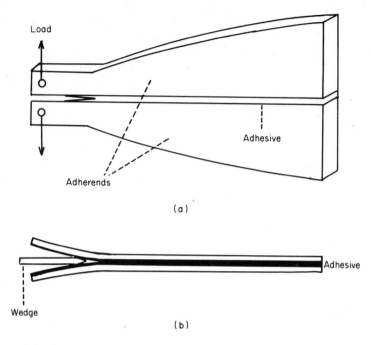

Fig. 1. (a) Thick adherend double cantilever beam (DCB) specimen. (b) Thin
double cantilever beam specimen (Boeing wedge test).

after the environmental exposure. However, McMillan[3] points out that
there was a satisfactory correlation with phenolic adhesives between
these tests and service performance, but this was not so when these
adhesives were replaced by epoxies. McMillan emphasises that in
addition to the residual strength of a joint, it is important to determine
the locus of failure (see Chapter 1).

As has been stated, because the working life of a bonded structure
may be 30 years in the case of an aircraft application, accelerated tests
are to some extent essential. A test may be accelerated by increasing
the temperature, load or humidity, or by a combination of these
factors. However, only moderate increases above the likely maximum
temperature should be used. It is especially important to avoid a shift
above the glass transition temperature, T_g, of the adhesive because the
mechanical properties will undergo large changes and also the uptake
of water will increase markedly (see section 5). McMillan[3] has pointed

out the danger of using loads much greater than would be experienced in practice. Such loads especially in combination with elevated temperatures might lead to creep rupture and hence to a misleading interpretation.

2.2.2. Unstressed Joints

Single, double or other types of joint are prepared and then subjected to various environments for periods up to a year or more. Typically an environment of 50°C and 100% R.H. would be used and joint strengths over a period of 10 000 h would be determined. Other typical tests involve salt sprays at room temperature and immersion in water at 40°C.

Although the use of unstressed joints may provide a useful screening test and an indication of the extent of corrosion,[4] it is advisable to carry out stressing tests to assess the suitability of a system for a structural application.

2.2.3. Stressed Joints

Various methods are available to examine the long-term durability of stressed joints. A common procedure is to use strings of lap joints which may be stressed by dead loads, springs or washers;[5] the joints are usually stressed to a selected percentage of the initial strength, typically 20%. The residual joint strength is measured after a given period of exposure to the chosen environment, e.g. 1000 h at 50°C and 100% R.H. Alternatively the time to failure at a given stress is determined (Fig. 2). Many systems show a stress endurance limit, i.e. at relatively low stresses there is no indication that failure will occur (Fig. 3). However, it is possible that failure would occur if the time of exposure was increased.

More recently double-cantilever beam (DCB) specimens (Fig. 1) have become widely used. This procedure was developed by Mostovoy and Ripling.[6,7] It is believed[4] that this type of test duplicates the characteristics of in-service failure most closely and it led to the introduction of the Boeing wedge test (Fig. 1). These self-stressing tests are generally much shorter in duration, usually involving hours rather than weeks or months; some typical wedge test results are given in Fig. 4 showing poor and good durability. The test is especially useful to assess the quality of surface pretreatments which is one of the most important factors affecting durability (section 4.3). In fact, Boeing use the wedge test as a quality control of the surface pretreatment.[8]

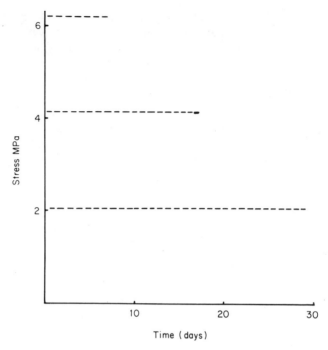

Fig. 2. Effect of different stress levels on the time to failure (at 45°C and 100% R.H.) for some lap joints involving 2024-T3 clad alloy and a modified epoxide adhesive (after ref. 8).

Bethune has described experiments on sections of service aircraft which show that the wedge test can be very discriminating.[8] Bonded panels with disbond in progress and sound panels were both subjected to wedge testing and the residual shear strengths were also determined. The residual shear strength always exceeded $27 \cdot 6$ MN m^{-2} (4000 psi), but the crack growth was much greater in the disbond regions.

As Minford describes in Chapter 4, Alcoa have used an alternative method[9] with a separate stressing ring for each joint (Fig. 5). Each ring is calibrated and can be accurately loaded. The procedure involves smaller environmental chambers than strings of joints, but by its nature requires individual calibration.

Although subjecting joints to a steady stress under a particular environment provides useful information, it is clear that such tests are not ideal for aircraft and other applications where the imposition of

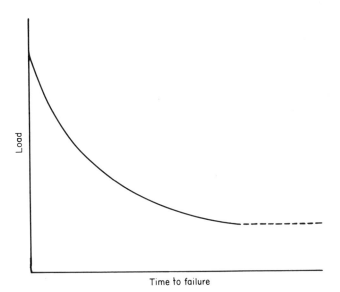

Fig. 3. Stress endurance limit: the dashed line represents the load at which there is no failure.

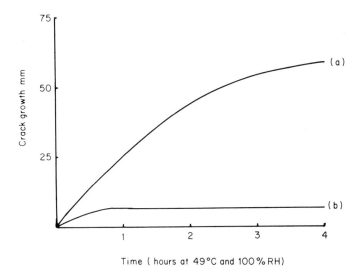

Fig. 4. Some typical Boeing wedge test results showing (a) poor durability with interfacial failure and (b) good durability.

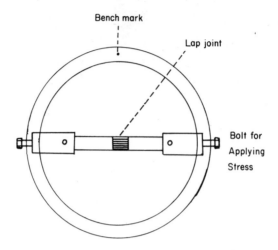

Fig. 5. Alcoa stressing device.

stress is intermittent. Cyclic testing is therefore becoming increasingly important. Frazier and Lajoie concluded[10] that loading for one hour followed by 15 min unloaded gave a good compromise cycle for a range of adhesives. McMillan[4] reported lap shear tests at three cyclic rates and noted that the slowest cyclic rate of 0·8 cph (one hour at maximum load, 15 min at minimum load) was more damaging in terms of damage per stress cycle than either 10 cph or 1800 cpm.

2.3. Outdoor Trials

Long-term outdoor durability trials are often carried out in conjunction with laboratory tests. In principle any type of adhesive joint is suitable, but lap joints are most commonly used. The environments chosen range from dry/hot (desert) to wet/hot (jungle) to industrial to seacoast locations.

Eickner described outdoor trials on aluminium joints carried out in the 1950s at a site close to the seashore at Miami, Florida and another site $\frac{1}{4}$ mile from the Panama Canal Zone.[11-13] The work was extended to include stressed joints.[14] In 1967, Wegman et al.[15] reported durability trials involving a Panama (moderate jungle) site, a temperate site at the Picatinny Arsenal, Dover, Delaware and a dry climate at Yuma, Arizona. There was a wide difference in the durability obtained with the 14 adhesives examined, but in general, the Panama site caused the most serious deterioration in joint strength.

TABLE 1
RELATIVE JOINT STRENGTH LOSS CAUSED BY DIFFERENT ENVIRONMENTS[3]

Mild flight service environment	Semi-tropical natural weather exposure	Ground–air–ground cycle	Condensing humidity elevated temperature	Severe flight service environment	5% salt spray exposure
<	<	<	<	≪	

Cotter[16] has described some more recent weathering trials carried out at three sites having widely different conditions. These sites were at the Royal Aircraft Establishment, Farnborough and at the Joint Tropical Research Unit (JTRU) sites at Innisfail (wet/hot) and Cloncurry (dry/hot) in Australia. The mean annual temperatures were 9, 24 and 26°C respectively with corresponding rainfalls of 0·48, 3·23 and 0·28 m. The average relative humidities of Innisfail and Cloncurry were 83 and 34% respectively. Some of the results of these trials are given in section 4.5. Normally, the durability deteriorates with increasing temperature and humidity, but anomalies do occur; for example, stressed joints involving a modified epoxide deteriorated more on a temperate site than at a wet/hot site.[16]

McMillan[3] has compared the severity of a semi-tropical natural environment with various laboratory tests and in-service conditions (Table 1).

3. ENVIRONMENTAL FACTORS AFFECTING DURABILITY

3.1. General
Many factors affect the durability of adhesive joints. Some, such as water ingress, are determined by the environment in which the bonded structure operates. Others, such as pretreatment or adhesive selection, are under the control of the manufacturer; these are discussed in section 4.

The main environmental factors likely to reduce durability are exposure to liquids or their vapours, ions, elevated temperatures and stress. Usually combinations of these factors will operate and this can be particularly damaging. Of the liquids likely to cause damage to joints, water is by far the most important, although other liquids such as antifreeze, hydrocarbon fuels and lubricating oils may also cause problems.

3.2. Water

Water may enter a bonded structure along an interface, by diffusion through an adhesive or by movement along cracks in the adhesive. After immersion in water or exposure to high humidity, a decrease in joint strength results, but the magnitude of the decrease depends on many factors including the time of exposure, temperature, alloy composition, pretreatment and the adhesive used.

Many studies on the effect of water on durability have been carried out. Figure 6 shows the effect of a wet/hot site on the joint strength of aluminium bonded with various adhesives, the substrates having been pretreated with chromic acid;[16] the loss in strength with the epoxide–polyamide is much greater than with most other structural adhesives (see section 4.5).

Some fairly typical results of laboratory tests on unstressed joints are given in Fig. 7. They show that the joint strengths of aluminium double

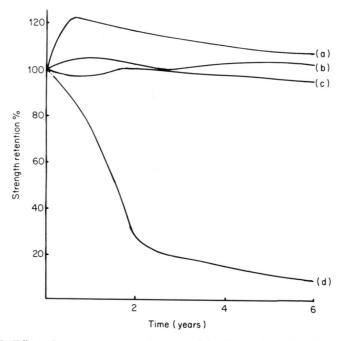

Fig. 6. *Effect of exposure to a wet/hot site of double overlap joints (unstressed) bonded with various adhesives:* (a) *epoxide–Novolac;* (b) *vinyl–phenolic;* (c) *nitrile–phenolic;* (d) *epoxide–polyamide.*

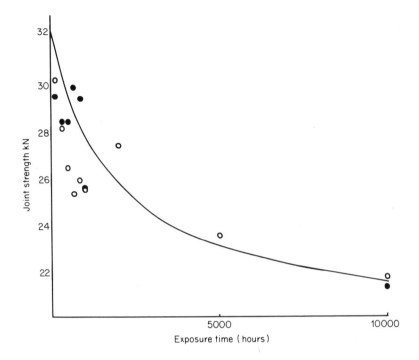

Fig. 7. Effect of 50°C and 100% R.H. on double overlap joints bonded with the epoxide BSL 312; ○, unstressed joints; ●, joints stressed to 20% of dry strength; the error bars for the 10 000 h exposure overlap (after ref. 26).

lap joints bonded with the epoxide BSL 312 (Ciba–Geigy) decrease to about 67% of the original value after 10 000 h (*ca* 60 weeks) exposure to 50°C and 100% R.H.

The reasons for the loss in joint strength due to water ingress have been the subject of much controversy and they are discussed in section 5 of this chapter and in Chapter 3. However, at this stage it is noted that at least for some systems, there is a linear relationship between the strength of aluminium lap joints and the fractional uptake of water.[17] Another important feature of durability studies that requires explanation is the recovery of joint strength on drying[18,19] and this is also discussed in section 5.

Although exposure of bonded structures to high humidities usually leads to substantial loss of strength, it has been observed that moderate levels of humidity do not lead to deterioration. Gledhill *et al.*[20] exposed epoxide-bonded joints to 55% R.H. at 20°C for 2500 h and found no weakening. Likewise, Moloney *et al.*[18] found no significant strength loss after exposing aluminium lap joints, bonded with the epoxides FM1000 (American Cyanamid) and BSL 312, for 10 000 h at about 20°C and 45% R.H.

Occasionally increases in joint strength have been observed after exposure to water. For example, Cotter[16] reported that unstressed aluminium joints bonded with an epoxide–Novolac adhesive gave increased strengths after exposure to the wet/hot conditions at Innisfail, Australia.

3.3. Temperature

Moderate temperatures by themselves do not normally have an adverse effect on structural bonds. For example, after the exposure of joints involving six different adhesives to a dry/hot site for six years,

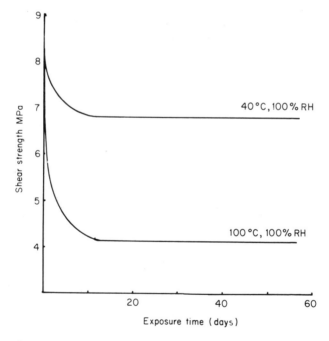

Fig. 8. *Effect of elevated temperature and 100% R.H. on epoxide/aluminium joints (after ref. 21).*

only one of the adhesive systems showed any significant deterioration and some showed an improvement.[16] Any loss in strength due to chemical degradation would require much higher temperatures. However, the combination of elevated temperatures and high humidities can have a serious effect on durability. Increasing temperature will increase the rate of uptake of water and accelerate chemical degradation. The results of Orman and Kerr[21] in Fig. 8 show that after a few days at 40°C and 100% R.H. the strengths of some aluminium/epoxide joints fell by about 15% and remained essentially constant thereafter for up to 50 days. At 100°C, the joints also reached a constant value after a few days, but at a much lower level.

Heating cured epoxides containing water to above 100°C can cause micro-cracks which would lead to increased water uptake,[22,23] and also act as stress raisers. There is therefore a danger that this would occur in epoxide and other structural adhesives in supersonic aircraft.

3.4. Stress

Joints may be subjected to intermittent or sustained stress which, combined with high humidity or elevated temperatures, can have a serious effect on the durability of aluminium joints.

Carter[24,25] reported that dramatic bond failures can occur in natural environments for epoxide-bonded aluminium joints if the stress is high. Carter reported that relative humidities as low as 50% reduce the durability of many structural adhesives under stress.

Some results described by Minford[9] demonstrate the critical role of stress. Figure 9 gives the unstressed durability results for an aluminium alloy, pretreated with chromic acid and bonded with two adhesives. After 700 days exposure to 51·7°C and 100% R.H., the joint strengths using a one-part vinyl–phenolic paste had decreased more than with a two-part room temperature cure adhesive. However, under conditions of applied stress, the vinyl–phenolic system gave much better durability (Fig. 10).

The environment has a very marked effect on the durability of stressed joints.[9] This is demonstrated by the results in Fig. 11 which show that the durability of etched aluminium joints, bonded with a nylon-modified epoxide, is much worse in an environment of 51·7°C and 100% R.H. than in water at room temperature; if the stress is $3·4 \, MN \, m^{-2}$ (500 psi), the time to failure is about 15 times greater under the latter conditions.

As well as by determining the time to failure at a given load, the

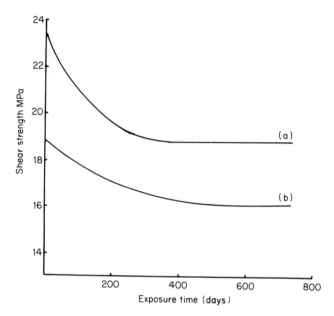

Fig. 9. *Durability tests at 51·7°C and 100% R.H. for unstressed joints involving acid-etched 6061-T6 alloy bonded with (a) one-part vinyl–phenolic paste cured at 176·7°C and (b) two-part room temperature-cure epoxide (after ref. 9).*

effect of stress may be examined by imposing a load, typically 20% of the initial joint strength, for a given time and then determining the residual joint strength. Stress does not always cause an additional reduction in joint strength. This has been shown[26] for the epoxide adhesive BSL 312 with and without carrier. The results for the adhesive with no carrier are shown in Fig. 7. However, for other systems, the effect of stress on residual strength is marked, as shown by the results in Fig. 12.[4]

McMillan[4] argues that stressing tests give failure similar to in-service bonds, i.e. interfacial, whereas unstressed joints give cohesive failure within the adhesive. McMillan emphasises the importance of fatigue testing. He notes that the combined effect of environmental exposure and slow cyclic rates can be especially damaging. Slow cycling is often encountered in aircraft.

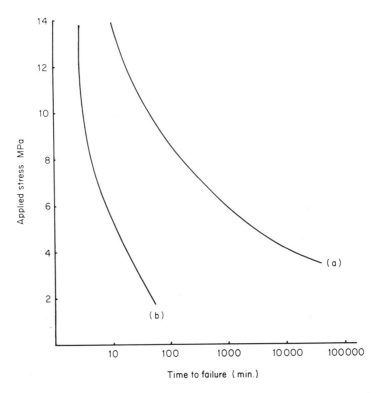

Fig. 10. *Durability tests at 51·7°C and 100% R.H. for stressed joints involving acid-etched 6061-T6 alloy bonded with (a) one-part vinyl–phenolic paste cured at 176·7°C and (b) two-part room temperature-cure epoxide (after ref. 9).*

3.5. Salt

Aircraft and road vehicles may be exposed to salt in coastal areas and cars will often be subjected to it on roads in the winter. Salt can have a very severe effect on joint strength durability. In one set of tests,[3] it was found that exposure to a 5% salt spray for three months had a more severe effect on aluminium joints (stressed DCB) than exposure to a semi-tropical environment for three years.

Table 2 gives some durability results for three alloys of different corrosion resistance.[9] The results clearly show that the effect of a salt

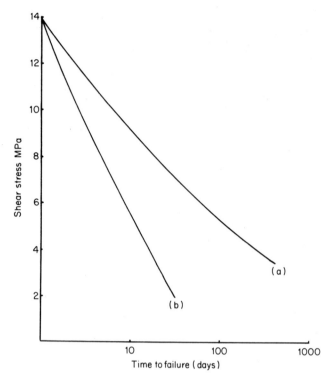

Fig. 11. Stressed aluminium (etched 6061-T6) joints bonded with a nylon-modified epoxide exposed to (a) room temperature water and (b) 51·7°C at 100% R.H. (after ref. 9).

spray is generally much more severe than exposure to high humidity even at 51·7°C. Deterioration in joint strength is normally much more rapid with salt solution than with water. This supports the view that, in the former case, corrosive attack along the bondline is involved, whereas attack by water first involves diffusion through the adhesive (Chapter 3).

4.   OTHER FACTORS AFFECTING DURABILITY

4.1. General

Section 3 concerned the environmental factors affecting durability. Other factors under the control of the manufacturer that also affect

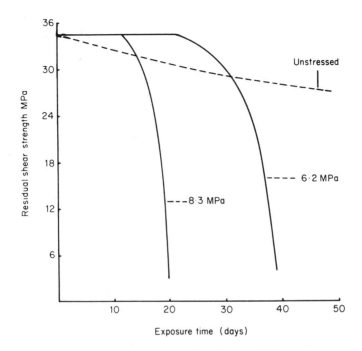

Fig. 12. Residual strengths of unstressed and stressed joints after exposure to 60°C and 100% R.H. (after ref. 4).

TABLE 2

EFFECT OF DIFFERENT ENVIRONMENTS ON THE DURABILITY OF VARIOUS ALLOYS BONDED WITH A ONE-PART EPOXIDE ADHESIVE[9]

Exposure conditions	Joint strengths $(MN\,m^{-2})^a$								
	Alloy 2036-T4			Alloy 6151-T4			Alloy X5085-H111		
	A	B	C	A	B	C	A	B	C
None	12·7	13·3	14·7	17·6	18·5	17·4	15·5	15·6	14·5
3 months at 23·9°C and 85% R.H.	2·6	2·2	15·2	9·3	10·7	16·6	12·8	13·0	13·2
3 months at 51·7°C and 100% R.H.	0·14	0	4·8	6·3	7·2	7·6	8·7	10·5	8·5
3 weeks 5% salt spray at 35°C	0	0	7·6	0·55	1·0	14·4	4·7	3·7	8·3

^{a} A, mill finish; B, vapour degreased; C, chromate–phosphate conversion coating.

durability, especially alloy composition, pretreatment, primer and adhesive, are discussed in this section.

4.2. Alloy Composition

Studies at Alcoa established a general relation between durability and the corrosion resistance of aluminium alloys.[9] The resistance of joints to seacoast environments or salt spray tests increased with enhanced corrosion resistance of the alloys. For the alloys shown in Table 2, the relative corrosion resistance is X5085>6151>2036. In general, the alloy X5085 gave the best durability, this being especially true when no pretreatment or only a vapour degrease was given.

The effect of magnesium in aluminium alloys on durability has been of considerable interest. Kinloch and Smart[27] noted a correlation between the magnesium concentration in the oxide layer and the durability of aluminium alloy joints. They used X-ray photoelectron spectroscopy (XPS) to determine the level of magnesium in the surfaces of a particular alloy after various pretreatments (see section 4.3). They found the level of magnesium present as magnesium oxide in the aluminium oxide layer to be in the order:

solvent-degreased > grit-blasted ≫ chromic acid-etched

> phosphoric acid-anodised

This observation is very significant because the order is the reverse of that obtained for durability. However, Kinloch and Smart emphasised that there were other important factors affecting durability. In a recent study, Kinloch et al.[28] have extended their work to three aluminium alloys containing 0–5 atomic % magnesium. The surface concentrations of magnesium in the three alloys were determined by XPS. For a given pretreatment, it was found that the durability decreased with increasing magnesium concentration in the surface, although the importance of other factors affecting durability was again emphasised.

Another factor affecting durability is the selection of clad or bare alloy. Cladding aluminium alloys with commercially pure aluminium was developed to improve their corrosion resistance. However, various publications have shown that clad alloys show durability inferior to that of bare alloys as far as salt spray tests are concerned.[29–31] This is in agreement with experience in Vietnam where aircraft were concentrated in coastal areas and on aircraft carriers.[9] Riel[30] proposed a mechanism involving galvanic action to explain the greater susceptibility of clad alloys to corrosive delamination. McMillan[3] has pointed out that the susceptibility of clad alloys to bondline corrosion is highly dependent on the composition of the base metal, the alloy 7075-T6 being much more susceptible than 2024-T3.

Although clad alloys are generally disfavoured in the USA,[31] European experience has shown that durable joints can be obtained provided a number of features are adopted.[16] These include the correct pretreatment, the use of a corrosion inhibiting primer and a suitable adhesive, sealed gluelines and the avoidance of mechanical fasteners through bonded areas.

Cotter and Kohler[1] reported that the thickness and density of the oxide layer after chromic acid anodising is dependent on the nature of the alloy, and also that clad alloys give thicker and smoother oxides.

4.3. Pretreatments

4.3.1. General

Many pretreatments for aluminium are available; these have recently been described and discussed by Moloney.[32] Degreasing or mechanical treatments such as grit blasting may be adequate in undemanding applications. However, if a joint is to be subjected to humid conditions, it is necessary to use an etching or anodising pretreatment to achieve good durability. The results in Table 3[33] show that poor durability is obtained with degreasing or grit blasting in a water immersion test at 40°C. The results in Table 2 show that vapour degreasing results in poor durability for joints exposed to humid conditions or to a salt spray. An exception has been noted by Minford;[9] grit-blasted aluminium gave good durability on a seacoast site.

TABLE 3

LOSS IN LAP SHEAR STRENGTH $(MN\ m^{-2})$ OF ALUMINIUM (BS 3L73) JOINTS AFTER IMMERSION IN WATER AT 40°C FOR 30 DAYS[33]

Temperature of adhesive cure (°C)	Pretreatment		
	Degreasing	Grit blasting	Etching
23	19·4	13·2	1·1
80	13·7	13·9	1·7

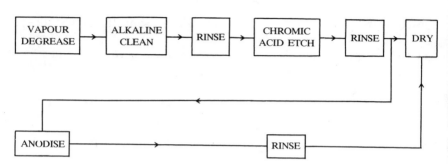

Fig. 13. Outline of chemical pretreatments for aluminium.

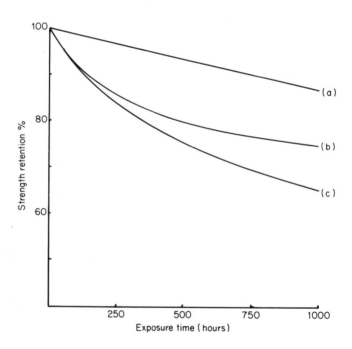

Exposure time (hours)

Fig. 14. Effect of pretreatment on the durability of aluminium single overlap joints bonded with a modified epoxide and immersed in deionised water at 50°C. (a) Phosphoric acid anodising; (b) chromic acid anodising; (c) chromic acid etched (after ref. 16).

The three methods usually used prior to bonding aluminium are:

(a) Chromic acid etching;
(b) Chromic acid anodising;
(c) Phosphoric acid anodising.

The steps involved in these pretreatments are outlined in Fig. 13; for further details see ref. 32. In general, the order of durability provided by the pretreatments is (a) < (b) < (c). Figure 14 gives the results comparing the three pretreatments in a water immersion test at 50°C and Fig. 15 compares the durability obtained with various pretreatments of stressed joints under hot humid conditions.[34] Anodising also gives superior results in salt spray tests and seacoast exposure.

The relative effectiveness of the three chemical pretreatments depends to some extent on the system examined and on the test used.

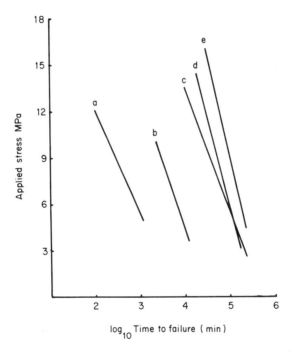

Fig. 15. *Effect of pretreatment on the durability of aluminium joints bonded with a nitrile–epoxide and subjected to applied stress at 52°C and 100% R.H. (a) Degreased; (b) sulphuric acid-anodised (unsealed); (c) chromic acid-etched; (d) chromic acid-anodised; (e) phosphoric acid-anodised (after ref. 34).*

D. M. Brewis

TABLE 4
STRENGTH RETENTION OF CHROMIC ACID-ANODISED ALUMINIUM LAP JOINTS, BONDED WITH BSL308A (Ciba–Geigy) AFTER EXPOSURE TO VARIOUS ENVIRONMENTS FOR 30 DAYS

Alloy	Rinse condition	Strength retention (%)				
		Initial strength	100% R.H.	Salt spray	Salt solution	Deionised water
L72 (clad)	Deionised water spray	100	81	84	3	85
	Cold tap water soak	86	78	78	a	80
	Hot tap water rinse	89	46	26	a	56
L70 (bare)	Deionised water spray	99	86	95	78	87
	Cold tap water soak	94	78	92	58	76
	Hot tap water rinse	81	61	71	40	69

[a] Specimens delaminated during exposure.

Phosphoric acid anodising generally gives superior durability (see for example ref. 30). However, Cotter and Kohler[1] have shown that if spray rinsing with cold deionised water is used instead of rinsing in hot tap water, then the durability results with chromic acid anodising were much improved (Table 4). The durability of joints immersed in salt solution at 50°C was then actually superior to phosphoric acid anodising, as were the results for two systems exposed to 50°C and high humidity (Table 5).

TABLE 5
EFFECT OF SURFACE PREPARATION ON CRACK PROPAGATION OF ANODISED ALUMINIUM CLEAVAGE SPECIMENS EXPOSED TO HIGH HUMIDITY AT 50°C[1]

Adhesive/primer	Anodised	Wedge crack extension $(mm)^a$				
		1h	3h	5h	24h	168h
BSL308A/108[b]	Chromic acid	1	2	2	2	2
BSL308A/108[b]	Phosphoric acid	2	3	3	4	4
FM73/127[c]	Chromic acid	0	0	1	1	1
FM73/127[c]	Phosphoric acid	1	2	2	2	2

[a] Initial crack lengths 55–61 mm for BSL308A, 49–55 mm for FM73. Crack extensions are the mean of five results. All specimens failed cohesively.
[b] Ciba–Geigy.
[c] American Cyanamid.

4.3.2. Chromic Acid Etching

During the etching process, the original oxide is removed, *ca* 1–3 μm of metal surface is dissolved and a new oxide layer is formed. Some workers have argued that the new oxide is formed during the rinsing stage, but work by Sun *et al.*[35] using Auger spectroscopy (AES; Chapter 2) has indicated that the formation of a thin oxide layer takes place during etching and the layer thickens during rinsing. There has also been much argument regarding the type of oxide formed.[16]

The etching conditions must be carefully controlled to give good results and a new chromic acid solution can give rise to poor durability.[4] McMillan has compared the durability provided by a traditional chromic acid etch with that provided by an 'optimised' etch. The latter involved the addition of aluminium and copper, and longer treatment times.[4] It had previously been shown that uniform etching is assisted by the presence of small quantities of aluminium[8] and copper.[8,36] However, excess quantities of aluminium[37] and copper[38] can cause unsatisfactory results. Other ions can lead to poor results, including iron above 1 g dm^{-3}, chloride ions[40] and fluoride ions.[4,40]

Much work has been carried out on the rinsing conditions. Rinsing in tap water is necessary to achieve satisfactory results, deionised water leading to low joint strengths. The poor results were ascribed to the formation of a thick, relatively weak oxide layer.[41] It has subsequently been shown by AES that the oxide layer formed after rinsing in deionised water is much thicker than when tap water is used.[42] It is also important that rinsing in tap water should be carried out immediately the aluminium has been removed from the etching bath to prevent conversion of Cr^{6+} to Cr^{3+}, which is detrimental to oxide stability.

The effects of some processing variables in chromic acid etching on durability have been reported by Scardino and Marceau.[29] The crack growth in a wedge test was much greater if the etching and rinsing conditions were incorrect (Fig. 16).

Correct chromic acid etching gives a surface which is covered by microscopic pits within larger pits.[43] To obtain better resolution than SEM, Chen *et al.*[40] used transmission electron microscopy (TEM) to obtain stereo pairs which give a three-dimensional view of the surface (Chapter 2). They achieved a resolution of 2–3 nm and deduced that 'whisker-like' structures protruded from the surface. They also found that incorrect treatment could give a relatively smooth surface without the protrusions; for example, the presence of fluoride ions during

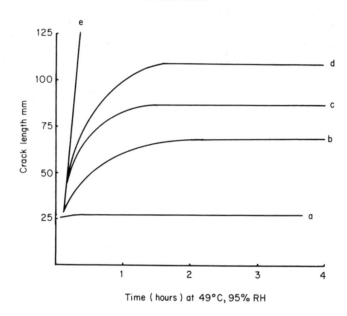

Fig. 16. *Effect of variables in chromic acid etching on durability as measured by the Boeing wedge test:* (a) *correct processing;* (b) *new solution;* (c) *etching solution 5·5°C below normal minimum temperature;* (d) *3 min rinse lag;* (e) *etching solution contaminated with 200 ppm F⁻ ions (after ref. 29).*

rinsing gives a much smoother surface. Venables et al.[44] have partially attributed the high initial joint strengths, obtained with a correct etch, to mechanical keying.

As noted above, chromic acid etching does not give as good durability as phosphoric acid anodising. An important reason is likely to be the instability of the oxide to water possessed by the etched material. This instability has been shown by Noland,[45] who used X-ray photoelectron spectroscopy (XPS; Chapter 2) to study etched aluminium before and after ageing for 1 h at 60°C and 100% R.H. The shift in the binding energy of the aluminium $2p$ peak indicates a change in the oxide and Noland attributed this to a relatively weak 'gelatinous boehmite' oxide.

Baun et al.,[46] using AES, observed a concentration of copper at the metal/oxide interface and noted significant changes in copper concentration with etching temperature and rinse times. Sun et al.[35] suggested that a high concentration of copper at the metal/oxide interface may have an important effect on durability. Using AES, they found up to 16 wt % copper with some alloys and attributed this to the diffusion of copper during etching. The presence of high levels of copper is known to assist in the electrochemical dissolution of the matrix in adverse environments.

4.3.3. Acid Anodising

Chromic acid and phosphoric acid anodising are widely used prior to the adhesive bonding of aluminium. Sulphuric acid anodising gives excellent corrosion protection, but bonding results have been poor either because of failure between the metal and the oxide[39] or because of failure within the oxide layer.[47]

Chromic acid anodising, which has been in use for many years, produces a thick dense oxide consisting of solid columns[44] which would be expected to give good corrosion resistance.

The Boeing company concluded that even 'optimised' etching with chromic acid was unsatisfactory for primary structural applications.[4] In the late 1960s Boeing commenced a programme to examine phosphoric acid anodising as a pretreatment and the method was introduced commercially in 1974.[4] Phosphoric acid anodising is less critically dependent than etching on processing variables such as the time between treatment and rinsing. It is also possible to use polarised light as a quality control test for the anodising pretreatment.[48]

Venables *et al.*,[44] using TEM in a scanning mode (Chapter 2), found the oxide layer was much thicker and the 'whiskers' longer with phosphoric acid anodising than for chromic acid etching. However, the thickness of the anodic oxide depends on the nature of the alloy.[49]

It has been known since the 1950s that phosphoric acid anodising produces surfaces that are more resistant to hydration than those produced with other anodising solutions including chromic acid.[50–52] Noland confirmed the high resistance to hydration of phosphoric acid-anodised aluminium by means of XPS, i.e. there was no significant shift in the binding energy of the $2p$ peak.[45]

It was suggested that the resistance to hydration was due to the presence of phosphate ions,[50,52] but it should be noted that the phosphorus present is concentrated near the surface of the oxide.[53–55] Also, various metallic elements may be present in the oxide layer or may be concentrated at the metal/oxide interface; both these occurrences can affect the stability of the interphase.[49] Sun *et al.*,[35] and

Kinloch *et al.*,[28] have shown that certain elements such as copper and magnesium are detrimental to oxide stability.

As described in Chapter 2, Davis *et al.*,[56] on the basis on XPS and electron microscopy data, concluded that the hydration of phosphoric acid-anodised aluminium occurs in three steps. The first stage is the reversible adsorption of water by the aluminium phosphate at the surface. The second step, which appears to be rate-determining, involves the slow dissolution of the phosphate followed by rapid hydration of the aluminium oxide to the oxyhydroxide boehmite. During this stage, 'extensive morphological changes occur as the boehmite fills the pore cells and bridges the whiskers of the original surface'. The third stage involves the nucleation and growth of the trihydroxide bayerite on top of the boehmite. The boehmite adheres weakly to the aluminium and hence the joint strength is reduced. Davis *et al.* found that the higher the phosphate concentration, the longer the time for hydration to occur.

Not only does phosphoric acid anodising generally give more durable joints than chromic acid etching, but in a programme involving 1250 wedge specimens for each pretreatment, anodising gave much more consistent results as measured by crack growth (Fig. 17).[4] Al-

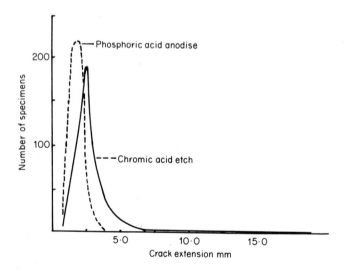

Fig. 17. *Consistency of the durability provided by phosphoric acid anodising and chromic acid etching as measured by the Boeing wedge test (after ref. 4).*

though etching often gave very good durability, the scatter of results was much greater than for anodising.

Study of the mode of failure is important when considering a pretreatment. Cohesive failure is associated with a successful pretreatment whereas apparent interfacial failure indicates an unsatisfactory performance. This is illustrated in some durability tests involving the modified epoxide adhesive BSL 312.[26] Initially joints involving etching and phosphoric acid-anodising treatments showed cohesive failure, but after exposure to hot humid conditions, they showed some apparent interfacial failure. This might be failure in the oxide layer, failure in the adhesive near the interface or true interfacial failure due to displacement of the adhesive by water (see section 5).

The reasons for the usual order of durability, i.e. degreasing < chromic acid etching < chromic acid anodising < phosphoric acid anodising, are not fully understood. There are large differences in the morphologies of the oxides.[8,44] However, the author supports the view that the oxide stability is paramount and that the highly porous structure produced by phosphoric acid anodising does not play an important role in durability. Nonetheless, the very thick oxide layer produced by chromic acid anodising[8,44] will protect the metal/oxide interphase from environmental attack; conversely the very thin oxide layer on degreased aluminium will provide little protection.

4.3.4. Other Methods

Although the three chemical methods discussed above, especially phosphoric acid anodising, provide very good durability, there is a need for simple methods where some applications are concerned. Phosphoric and chromic acid anodising involve seven stages (Fig. 13). It is also highly desirable to avoid the use of chromium compounds because of their toxicity and related disposal problems.

Smith[57] described a two-stage method to pretreat aluminium involving relatively harmless materials. The aluminium is degreased with a mixture of a solvent and a proprietary degreasing liquid, followed by washing in tap water or solutions of carbonates or bicarbonates at 80°C. It was claimed that the durability, as measured by the wedge test, was in some cases as good as that produced by phosphoric acid anodising. However, this claim does not appear to have been substantiated.

US Patent 4212701 describes a method involving sulphuric acid, ferric sulphate and water to pretreat aluminium.[58] Durability superior to that from chromic acid etching was indicated by wedge tests and by stressing lap joints at 60°C and 95% R.H.

A recent method described by Pocius and Claus[59] involves the use of a proprietary abrasion technique as an alternative to chromic acid etching prior to phosphoric acid anodising. The technique apparently only removes surface contaminants and a small amount of substrate. It is claimed that the durability, as measured by the wedge test, is as good as when chromic acid is used prior to anodising. As the method is simpler and less hazardous than using chromic acid, it is clearly worthy of further study.

4.4. Primers

A freshly prepared aluminium surface is highly reactive and will rapidly undergo interaction with water or organic contaminants in the atmosphere. Therefore, to achieve maximum interaction with the adhesive, it is necessary to bond the aluminium within a few hours of pretreatment or alternatively to prime the substrate. Priming is an operation whereby an organic coating is applied to the substrate by spraying, dipping or brushing. An important function of a primer is to permit storage of the pretreated part. It has been found that bondability is maintained for a year or more provided gross contamination is avoided.[60]

In addition, a primer may offer a number of other advantages. A primer is frequently of similar chemical structure to the adhesive, e.g. epoxide primers are often used prior to epoxide adhesives. Because the primer can be applied in a solution with a much lower viscosity than the thermosetting adhesive, better wetting of the substrate can be achieved. This improved wetting should not only give higher initial joint strengths, but will lead to fewer voids which could fill with water. Such pockets of water could in turn lead to weakening of the oxide or corrosion at the metal/oxide interface.

Durability can also be improved by the incorporation in the primer of corrosion-inhibiting agents such as strontium chromate. Corrosion-inhibiting primers are always used with bonded components in US Air Force aircraft.[60]

Sell[61] has pointed out that durability increases with the thickness of the primer layer up to the point where the primer acts as a separate layer. Kinloch[34] has also pointed out that the primer layer may become the weakest part of a joint.

Primers are normally applied from organic solvents. Water-based primers would eliminate toxicity and flammability problems associated with most organic solvents, and could have cost advantages. Reinhart[60] has reported the use of primers based on water-soluble polymers such as phenol-formaldehyde, resorcinol–formaldehyde, urea–formaldehyde, blends of these and also special epoxide resins. Initial durability studies involving wedge and residual strength tests showed that although these primers were inferior to a commercial product, the results were encouraging and Reinhart considered that a thinner and tougher film would lead to an improvement. He also described some promising work on the use of organic chromate primers.

The work described by Scardino and Marceau[29] illustrates the importance of a corrosion-inhibiting primer when chromic acid etching or anodising are used. In salt spray wedge tests, the crack growths were much greater with chromic acid etching or anodising if no corrosion-inhibiting primer was used. With phosphoric acid anodising the crack growth was small whether or not a primer was used; when a primer was used, a reduction in crack growth occurred with a clad alloy, but with a bare alloy the growth was actually greater. Bethune[8] described the effect of a corrosion-inhibiting primer on some stressed lap joints exposed to 78°C and 100% R.H. At a stress of $2 \cdot 1 \, MN \, m^{-2}$ (300 psi) the primer extended the time to fail from 28 to more than 250 days, while at a stress of $6 \cdot 2 \, MN \, m^{-2}$ (900 psi), the time to failure increased from about seven to about 40 days. However, the mode of failure remained apparently interfacial.

Silanes represent a special type of primer which ideally acts as a coupling agent between a substrate and the adhesive, i.e. it forms primary bonds with both materials. Gettings and Kinloch[62,63] found that a γ-glycidoxypropyltrimethoxysilane primer considerably improved the durability of steel joints whereas other silanes did not. Using secondary ion mass spectroscopy (SIMS; Chapter 2), they detected $FeSiO^+$ radicals with mild steel, and $FeSiO^+$ and $CrSiO^+$ radicals with stainless steel when using the glycidoxypropyltrimethoxysilane, but not with the other silanes (Chapter 1, Fig. 7). This is very strong evidence in favour of chemical bonding with the particular silane although it is not conclusive because it is possible that the radicals could be formed during the SIMS analysis.[64] Silanes may also improve the durability of aluminium/epoxide joints.[9,65]

Brockmann[66] found that priming aluminium with 8-hydroxyquinoline or 2,4,5,7-tetrahydroxyflavonene considerably increased the durability of joints bonded with phenolic or nitrile–epoxide adhesives.

4.5. Nature of Adhesive

The effect of the adhesive on durability is discussed in detail in Chapter 4, but some important results are summarised in this chapter.

Sell[61] has ranked adhesives in the order of durability that they provide with aluminium adherends as follows: Nitrile phenolics > high temperature epoxides > 121°C (250°F) curing epoxides > 121°C (250°F) curing rubber modified epoxides > vinyl epoxides > two-part room temperature curing pastes with amine cure > two-part room temperature curing pastes with anhydride cure > two-part polyurethanes.

Bethune[8] has reported the successful application of nitrile–phenolics in several models of aircraft. In contrast, a room temperature curing epoxide showed a high incidence of failures and a 121°C curing epoxide showed sporadic failure.

The results[16] in Figs 6 and 18 show the effect of adhesive on the durability at a hot/wet site (the average temperature and relative

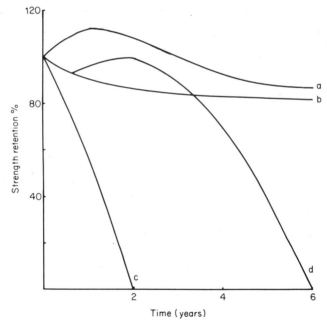

Fig. 18. *Effect of adhesive on strength retention of stressed aluminium double overlap joints at hot/wet site, Innisfail, Australia:* (a) *epoxide–Novolac;* (b) *nitrile–phenolic;* (c) *epoxide–polyamide;* (d) *vinyl–phenolic (after ref. 16).*

humidity being 24°C and 83%) of unstressed and stressed aluminium joints respectively. The tests involved clad alloy which had been etched in chromic acid. In the tests involving unstressed joints, three of the adhesives (epoxide–Novolac, nitrile–phenolic and vinyl–phenolic) showed excellent durability whereas the epoxide–polyamide showed a drastic loss of strength. On a hot/dry site (the average temperature and relative humidity being 26°C and 34%), the epoxide–polyamide showed a loss in joint strength of only about 10% in six years.

The application of stress under hot/humid conditions shows some marked differences between adhesives (Fig. 18). Again the epoxide–Novolac and nitrile–phenolic show superior durability, but the vinyl–phenolic shows rapid loss of strength after two years.

Generally, the higher the cure temperature of the adhesive the greater is the durability of the adhesive joint. When a 121°C (250°F) cure adhesive has been cured at a higher temperature, improved durability is obtained.[9]

Fabrics known as carriers are often contained in film adhesives to improve the ease of handling and to control glueline thickness. It might be expected that these carriers, which are usually polyester or nylon, might result in an increased uptake of water by wicking, with a resultant adverse effect on durability. However, Brewis *et al.*[26] showed that the incorporation of polyester and nylon carriers into the modified epoxide adhesive BSL 312 had a negligible effect on the durability of aluminium joints exposed to air at 50°C and 100% R.H. for periods up to 10 000 h.

Most of the studies on the durability of aluminium joints have involved epoxide adhesives. However, there is an increasing interest in the use of acrylic adhesives because of their rapid cure at room temperature and their tolerance to contaminants. These properties are particularly useful in non-aerospace applications where production speed and simplicity are more important than achieving the maximum possible performance.

The potential use of acrylics for bonding aluminium has been discussed by Zalucha.[67] He reported durability tests involving unprepared 6061-T6 aluminium. After 1000 h exposure to 55°C and 100% R.H., the joints retained 80–90% of the initial strength which compares favourably with joints involving aluminium etched with chromic acid. Stressed to 25% of their ultimate failing strength, untreated joints survived several days at 55°C and 100% R.H.; the failure indicated by ISS/SIMS was within the oxide layer. In unstressed tests with phosphoric acid-anodised clad alloy, the joints retained about 95% of their original strength after exposure to 60°C and 100% R.H. for 1000 h.

Lees[68] reported that good durability was obtained with the toughened acrylic Flexon 241 (Permabond). Joints involving etched aluminium and this adhesive decreased in strength by only 3% after 1000 h exposure to 45°C and 100% R.H.

4.6. Contamination

Contamination by organic or inorganic species can reduce durability. The contamination may occur before, during or after pretreatment. The effective removal of organic contaminants prior to etching is a routine procedure in the aircraft industry. Contamination of pretreated aluminium surfaces by organic molecules is minimised by priming (section 4.4).

Contamination by inorganic species is in some respects more difficult to control. Contamination of pretreatment baths was noted in section 4.3, but other contamination can occur during processing of the aluminium alloy and also from the adhesive. For example, Baun[64] has pointed out the possible sources of a considerable number of impurities during the processing of aluminium. The impurities include chlorine, magnesium, sodium and potassium which can all have a detrimental effect on durability.

5.   MECHANISM OF STRENGTH LOSS DUE TO WATER

The durability of aluminium joints is obviously a highly complex subject. A satisfactory understanding of the loss in joint strength must explain the effects of the factors mentioned in sections 3 and 4. A large amount of work has been carried out on viscous aspects of durability. However, it is often difficult to compare the results of different workers because of the many variables involved. Also, on many occasions, insufficient practical detail is published to permit a detailed analysis of the results.

Water is by far the most common agent to reduce the joint strength of aluminium bonds. The ingress of water into an adhesive joint has been discussed in detail in Chapter 3 and in ref. 17. Water can enter a joint by diffusion through the adhesive, by capillary action along cracks in the adhesive, by wicking along a fabric carrier or by transport along

the oxide/polymer interface. It has been shown that a linear .
exists between joint strength and water uptake for an epoxide resin
cured with di(1-aminopropyl-3-ethoxy) ether,[69] and for the epoxide
adhesive BSL 312;[26] there was normally a decrease in joint strength,
but in one case an increase occurred.

Although immersion of aluminium joints in water or exposure to
high humidity normally leads to a substantial loss in joint strength, it
has been observed by several groups of workers that exposure of joints
to moderate humidities for long periods does not normally lead to
strength loss. For example. DeLollis[70] reported that aluminium/epoxide
joints suffered no loss in strength after exposure to a laboratory environment for 11 years. Moloney *et al.*[18] found no significant strength
loss for aluminium joints bonded with FM 1000 and BSL 312, after
10 000 hours at *ca* 20°C and 45% R.H.

Such observations led Kinloch[34] to propose his idea of a critical
concentration of water in an adhesive joint, i.e. until the amount of
water in a joint exceeds a certain level, no loss of strength occurs. This
concept was developed by Gledhill *et al.*[20] They argued that at the
point where the critical concentration is exceeded, there will be a
relatively sharp boundary between the weakened regions of the joint
and the inner regions not affected by water ingress. The extent of
interfacial debonding is equivalent to a crack. Using fracture
mechanics, they predicted the fracture stress of some joints involving
an epoxide adhesive under different environmental conditions. Good
agreement with the experimental values was obtained (see Chapter 7,
Fig. 12).

Once the water concentration in a joint has exceeded the critical
value, there are several possible mechanisms by which joints can be
weakened, namely:

(a) Effect on the bulk properties of the adhesive.
(b) Displacement of adhesive from the substrate.[21,65,71,72]
(c) Weakening of the aluminium oxide.
(d) Reduction of ion-pair interaction across the oxide/polymer
     interface.[17,73]

The effect of water on the adhesive is discussed in detail in Chapter 3.
Water may plasticise the adhesive and this has been shown[74] to be the
main mechanism of strength loss with FM 1000. On exposure of
aluminium lap joints bonded with FM 1000 for 2500 h to 50°C and
100% R.H., the joint strength fell by about 45%. However, the mode

of failure remained largely cohesive. It is true that FM 1000 is abnormal in the high level of uptake of water, but other examples of adhesive plasticisation have been noted elsewhere.[3] Plasticisation is a reversible effect, but water may also cause irreversible changes, in particular hydrolysis, cracking or crazing. The conditions used to demonstrate hydrolysis in epoxides were severe[75,76] and the relationship with practical adhesive joints is not clear. Apicella *et al.*[77] observed an abnormal uptake of water and attributed this to microcavities. Not only can such cavities lead to an increase in water uptake, but they can also act as stress raisers and thereby reduce joint strength.

Kerr *et al.*[78] proposed that the adverse effect of water on aluminium was due at least in part to 'interference' with the metal oxide–polymer bonding. Gledhill and Kinloch[71] proposed a mechanism to explain how water could displace an adhesive from aluminium[65] and iron[71] oxides. Using a thermodynamic approach, they calculated that an aluminium oxide/epoxide interface would be stable when dry (work of adhesion = $+232 \text{ mJ m}^{-2}$), but if there is a layer of water at the interface, the work of adhesion has a negative value, namely -255 mJ m^{-2}, i.e. the oxide/epoxide interface is unstable in the presence of water. This mechanism explains the progressive reduction in joint strength with time and the change in the mode of failure from cohesive to interfacial. However, it does not explain why the joint strength falls to a definite level, say 70% of the initial joint strength, and then remains at that level for a long period. Nor is the mechanism consistent with the fact that joints can recover much of their lost strength on drying.

The mechanism of strength loss which has received the most support is the weakening of the oxide by hydration; for example, see refs 4 and 33 and Chapter 2. One strong piece of evidence in favour of this mechanism is provided by the work of Noland.[45] He showed that aluminium etched with chromic acid is much more susceptible to hydration than aluminium that has been anodised by phosphoric acid; it has been clearly established that the latter pretreatment imparts much better durability to aluminium joints (see section 4.3). Further support for this mechanism is the improvement in durability provided by corrosion-inhibiting primers (see section 4.4). The adverse effect of magnesium and copper can be best explained by a weakening of the oxide layer.[28,49,53] Bethune[8] used time to failure tests for some chromic acid-etched aluminium (Fig. 2) to support the proposal that water weakened the oxide by hydration. He argued that in 20 days the oxide had weakened to the extent that it could no longer sustain the

8·3 MN m^{-2} (1200 psi) applied stress, in 40 days it could no longer support 6·2 MN m^{-2} (900 psi) and in 80 days it could no longer support 4·1 MN m^{-2} (600 psi). Further support came from the fact that failure occurred within the oxide layer. In addition to a weakening of the oxide, it is possible that the adhesion at the metal/oxide interface is reduced.[79]

However, Kollek and Brockmann[80] found with chromic acid etching that failure usually occurs in the adhesive near the interface. They argue that fracture of the adhesive is of critical importance in understanding durability. They further propose that the nature of the pretreated surface is important because it affects the curing of the adhesive and also the interaction across the interface. They found that the degree of chemisorption was much greater with anodised than with etched surfaces.

Comyn proposed[17,73] that water can reduce the interaction between ion-pairs across an interface resulting in a decrease of joint strength. He explains how ions can be formed during the curing of an epoxide and how these ions would interact with the ionic oxide (see Chapter 3). The force of interaction between the ions is inversely proportional to the relative permittivity of the medium. Therefore water at the interface will reduce the relative permittivity and hence the attraction between the ions. The theory thus provides an explanation of the retention of a definite percentage of this initial joint strength on exposure to water for long periods and it explains the recovery of joint strength on drying. However, at this stage there is no direct evidence for the theory.

It is probable that for some systems, i.e. particular combinations of alloy–pretreatment–primer–adhesive, one mechanism may predominate, whereas for other systems two mechanisms may be important. Some mechanisms, e.g. plasticisation of the adhesive, are reversible while others, e.g. displacement by water of an adhesive from the substrate, are permanent.

The recovery of joint strength on drying gives further indirect evidence for the mechanism of strength loss due to water. Brewis *et al.*[26] examined the recovery of joint strengths involving the epoxide adhesive BSL 312. After 1000 h exposure to 50°C and 100% R.H., specimens etched with chromic acid retained 75% of their initial strength, whereas those involving phosphoric acid anodising retained 94%. After 1000 h drying at 50°C, the etched samples had recovered to 83% of the initial strength and the anodised samples had recovered completely. These results indicate that a permanent weakening of the oxide formed on etching had occurred,[81] and that only the strength loss due to plasticisation of the adhesive was recoverable. These results support the view that the anodised oxide is stable and the loss of strength is due to plasticisation by water. However, it is not possible to draw firm conclusions from these limited data, especially as the time scale is short in terms of durability tests.

It is important to examine the interrelation between the various factors affecting durability. For example, it was found that the strength of some aluminium joints bonded with DGEBA/1,3-diaminobenzene only deteriorated by 28% after exposure for 10 000 h to 50°C and 100% R.H. despite the fact that only a mechanical pretreatment was used.[82] It is believed that the better-than-expected durability is due at least in part to the relatively low rate of uptake of water by the adhesive. This means that loss of strength due to plasticisation will be low and also degradation of the interface will be reduced. It is interesting that there is only a modest recovery of joint strength on prolonged drying. This indicates that most of the damage when it does occur is due to a permanent change, probably a weakening of the interface or oxide.

The author believes that the balance of evidence of the many publications on the durability of aluminium joints, favours the view that the two predominant mechanisms of strength loss are plasticisation, which is reversible, and a permanent weakening of the aluminium oxide layer. The relative importance of the two mechanisms will vary with different systems. At one extreme would be the bonding of phosphoric acid-anodised aluminium with FM 1000, where most of the strength loss would be due to plasticisation. At the other extreme would be the bonding of mechanically pretreated aluminium with DGEBA/1,3-diaminobenzene. However, much more work will be necessary before a full understanding of durability is achieved. Work directed at the recovery of joint strength on drying and also the examination of the true locus of joint failure should be especially useful. It is clearly of paramount importance to know whether failure occurs at the metal/oxide interface, within the oxide layer, at the oxide/adhesive interface or within the adhesive.

6. CONCLUSIONS

(1) Much progress has been made in recent years regarding the durability of aluminium joints; however, we are still far from a

full understanding of the effects of alloy composition, pretreatment, primers and adhesives. The fairly recent introduction of a range of powerful surface analytical techniques should continue to advance the knowledge of these effects.

(2) The two substances which have the most serious effect on durability are salt and water. The former acts by corrosive attack along the bondline. Attack by water first involves diffusion of water through the adhesive, followed by one or more of several strength loss mechanisms. The author believes that plasticisation and weakening of the aluminium oxide are of particular importance.

(3) Phosphoric acid anodising normally provides the most durable joints, although a modified rinsing procedure with chromic acid anodising appears to give equally good results. From the practical viewpoint, phosphoric acid has the advantages of lower toxicity, lower susceptibility to processing variables and easier quality control. However, anodising pretreatments are complex and there is a need for simplifications, especially for applications other than in the aerospace industries.

REFERENCES

1. Cotter, J. L. and Kohler, R., *Int. J. Adhesion Adhesives*, **1** (1980), 23.
2. Reynolds, B. L., *Proc. Army Mater. Conf.*, **4** (1976), 605.
3. McMillan, J. C., *Developments in Adhesives—2*, ed. A. J. Kinloch, Applied Science Publishers, London (1981), p. 243.
4. McMillan, J. C., *Bonded Joints and Preparation for Bonding*, NATO AGARD Lecture Series 102 (March 1979), paper 7.
5. Brewis, D. M., Comyn, J., Cope, B. C. and Moloney, A. C., *Lab. Practice*, **28**(7) (1979), 743.
6. Mostovoy, S. and Ripling, E. J., *J. Appl. Polym. Sci.*, **13** (1969), 1083.
7. Mostovoy, S. and Ripling, E. J., *J. Appl. Polym. Sci.*, **15** (1971), 641.
8. Bethune, A. W., *SAMPE J.*, **11** (1975), 4.
9. Minford, J. D., *Treatise on Adhesion and Adhesives*, Vol. 5, ed. R. L. Patrick, Marcel Dekker, New York (1981), p. 45.
10. Frazier, T. B. and Lajoie, A. D., Durability of adhesive bonded joints, *Technical Report AFML-TR-74-26*, Bell Helicopter Company (March 1974).
11. Eickner, H. W., Weathering of adhesive bonded lap joints of clad aluminium alloys, *WADC Technical Report 54–447*, Part I, Forest Products Laboratory, (1955).
12. Eickner, H. W., Weathering of adhesive bonded lap joints of clad aluminium alloys, *WADC Technical Report 54–447*, Part II, Forest Products Laboratory, (1957).

13. Eickner, H. W., Weathering of adhesive bonded lap joints of clad aluminium alloys, *WADC Technical Report 54–447*, Part III, Forest Products Laboratory (1958).
14. Eickner, H. W., Environmental exposure of adhesive bonded metal lap joints, *WADC Technical Report 59–567*, Part I, Forest Products Laboratory (February 1960).
15. Wegman, R. F., Bodnar W. M., Duda E. S. and Bodnar M. J., *Adhesives Age* (October 1967), 22.
16. Cotter, J. L., *Developments in Adhesives—1*, ed. W. C. Wake, Applied Science Publishers, London (1977), p. 1.
17. Comyn, J., *Developments in Adhesives—2*, ed. A. J. Kinloch, Applied Science Publishers, London (1981), p. 279.
18. Moloney, A. C., Brewis, D. M., Comyn, J. and Cope, B. C., *Adhesion 5*, ed. K. W. Allen, Applied Science Publishers, London (1981), p. 133.
19. Brewis, D. M., Comyn, J., Cope, B. C. and Moloney, A. C., *Polymer*, **21** (1980), 344.
20. Gledhill, R. A., Kinloch, A. J. and Shaw, S. J., *J. Adhesion*, **11** (1980), 3.
21. Orman, S. and Kerr, C., *Aspects of Adhesion 6*, ed. D. J. Alner, University of London Press (1971), p. 64.
22. Browning, C. E., *Polym. Eng. Sci.*, **18**(1) (1978), 16.
23. Browning, C. E., *Nat. SAMPE Symp. Exhib.*, **22** (1977), 365.
24. Carter, G. F., *Adhesives Age* (October 1967), 32.
25. Carter, G. F., Durability of adhesive peel joints when stressed in water, *ASTM STP 401* (August 1966), p. 28.
26. Brewis, D. M., Comyn, J., Cope, B. C. and Moloney, A. C., *Polym. Eng. Sci.*, **21**(12) (1981), 797.
27. Kinloch, A. J. and Smart, N. R., *J. Adhesion*, **12** (1981), 23.
28. Kinloch, A. J., Bishop, H. E. and Smart, N. R., *J. Adhesion* **14** (1982), 105.
29. Scardino, W. and Marceau, J. A., *J. Appl. Polym. Sci., Appl. Polym. Symp.*, **32** (1977), 51.
30. Riel, F. J., *SAMPE J.*, **7** (1971), 16.
31. Rogers, N. L., *Proc. 5th. Nat. Tech. Conf., SAMPE, Kiemesha Lake, New York* (1973), p. 160.
32. Moloney, A. C., *Surface Analysis and Pretreatment of Plastics and Metals*, ed. D. M. Brewis, Applied Science Publishers, London (1982), p. 175.
33. Garnish, E. W., *Adhesion 2*, ed. K. W. Allen, Applied Science Publishers, London (1978), p. 35.
34. Kinloch, A. J., *J. Adhesion*, **10** (1979), 193.
35. Sun, T. S., Chen, J. M., Venables, J. D. and Hopping, R., *Applns. Surf. Sci.*, **1** (1978), 202.
36. Smith, A. W., *J. Electrochem. Sci.*, **120** (1973), 1551.
37. Bijlmer, P. F. A., *Metal Finishing* (May 1975), 47.
38. Bijlmer, P. F. A., *Metal Finishing* (December 1971), 34.
39. Reinhart, T. J., *Proc. Army Mater. Conf.*, **4** (1976), 201.
40. Chen, J. M., Sun, T. S. and Venables, J. D., *Nat. SAMPE Symp. Exhib.*, **22** (1978), 25.
41. McCarvill, W. T. and Bell, J. P., *J. Appl. Polym. Sci.*, **18** (1974), 343.

42. Pattnaik, A. and Meakin, J. D., *J. Appl. Polym. Sci., Appl. Polym. Symp.*, **32** (1977), 145.
43. Bijlmer, P. F. A., *J. Adhesion*, **5** (1973), 319.
44. Venables, J. D., McNamara, D. K., Chen, J. M., Sun, T. S. and Hopping, R., *Nat. SAMPE Tech. Conf.*, **10** (1978), 362.
45. Noland, J. S., *Adhesion Science and Technology 9A*, ed L. H. Lee, Plenum, New York (1975), p. 413.
46. Baun, W. L., McDevitt, N. T. and Solomon, J. S., *ASTM Spec. Tech. Publ.*, **596** (1976), 86.
47. Bijlmer, P. F. A., *Metal Finishing* (April 1972), 30.
48. McMillan, J. C., Quinlivan, J. T. and Davis, R. A., *SAMPE Q.* (April 1976), 13.
49. McDevitt, N. T. and Solomon, J. S., Thin anodic oxide films on aluminium alloys and their role in the durability of adhesive bonds, *Technical Report AFML-TR-79-4216* (February 1980).
50. Hunter, M. S., Towner, P. F. and Robinson, D. L., *Proc. Amer. Electroplating Soc.*, **46** (1959), 220.
51. Keller, F., Hunter, M. S. and Robinson, D. L., *J. Electrochem. Soc.*, **100** (1953), 411.
52. Amore, C. J. and Murphy, J. F., *Metal Finishing*, **63** (1965), 50.
53. Kinloch, A. J., *Adhesion 6*, ed. K. W. Allen, Applied Science Publishers, London (1982), p. 95.
54. Solomon, J. S., Hanlin, D. and McDevitt, N. T., *Adhesion and Adsorption of Polymers*, ed. L. H. Lee, Plenum, New York (1980), p. 103.
55. Sun, T. S., McNamara, D. K., Ahearn, J. S., Chen, J. M., Ditchek, B. and Venables, J. D., *Appl. Surf. Sci.*, **5** (1980), 406.
56. Davis, G. D., Sun, T. S., Ahearn, J. S. and Venables, J. D., *J. Mater. Sci.*, **17** (1982), 1807.
57. Smith, T., *J. Adhesion*, **9** (1977), 313.
58. US Patent 4 212 701 (15 July 1980).
59. Pocius, A. V. and Claus, J. J., *SAMPE Meeting, Munich* (Noember 1981).
60. Reinhart, T. J., *Adhesion 2*, ed. K. W. Allen, Applied Science Publishers, London (1978), p. 87.
61. Sell, W. D., Some analytical techniques for durability testing of structural adhesives, *19th Nat. SAMPE Symp.* (April 1974).
62. Gettings, M. and Kinloch, A. J., *J. Mater. Sci.*, **12** (1977), 2511.
63. Gettings, M. and Kinloch, A. J., *Surf. Interface Anal.*, **1** (1979), 189.
64. Baun, W. L., *Surface Analysis and Pretreatment of Plastics and Metals*, ed. D. M. Brewis, Applied Science Publishers, London (1982), p. 45.
65. Kinloch, A. J., Dukes, W. A. and Gledhill, R. A., *Adhesion Science and Technology 9B*, ed. L. H. Lee, Plenum, New York (1975), p. 597.
66. Brockmann, W., *Adhesives Age* (June 1977), 30.
67. Zalucha, D. J., *13th Nat. SAMPE Tech. Conf.* (October 1981), p. 92.
68. Lees, W. A., *Advances in Adhesives. Applications, Materials and Safety*, eds. D. M. Brewis and J. Comyn, Warwick Publishing, Birmingham, UK (1982), Chapter 6.
69. Brewis, D. M., Comyn, J. and Tegg, J. L., *Int. J. Adhesion Adhesives*, **1** (1980), 35.

70. DeLollis, N. J., *Nat. SAMPE Symp. Exhib.*, **22** (1977), 673.
71. Gledhill, R. A. and Kinloch, A. J., *J. Adhesion*, **6** (1974), 315.
72. Butt, R. I. and Cotter, J. L., *J. Adhesion*, **8** (1976), 11.
73. Comyn, J., *Adhesion 6*, ed. K. W. Allen, Applied Science Publishers, London (1982), p. 159.
74. Brewis, D. M., Comyn, J. and Shalash, R. J. A., *Int. J. Adhesion Adhesives*, (October 1982), 215.
75. Antoon, M. K. and Koenig, J. L., *J. Polym. Sci., Physics Edn.*, **19** (1981), 197.
76. Antoon, M. K., Koenig, J. L. and Serafini, T., *J. Polym. Sci., Physics Edn.*, **19** (1981), 1567.
77. Apicella, A., Nicolais, L., Astanta, G. and Drioli, E., *Polym. Eng. Sci.*, **21** (1981), 17.
78. Kerr, C., MacDonald, N. C. and Orman, S., *J. Appl. Chem.* **17** (1967), 62.
79. Venables, J. D., *20th Conf. on Adhesion and Adhesives, City University, London* (April 1982).
80. Kollek, H. and Brockmann, W., Techniques of measuring the interfacial ageing mechanisms in metal bonds, *26th National SAMPE Symp.* (April 1981).
81. Eley, D. D. and Rudham, R. *Adhesion Fundamentals and Practice*, ed. UK Min. of Tech., Elsevier, London (1969), p. 91.
82. Raval, A. K., private communication.

APPENDIX
ALLOY COMPOSITIONS
(wt %, the remainder being aluminium)

Alloy	Si	Fe	Cu	Mn	Mg	Zn	Cr
2024	0·50	0·50	3·8–4·9	0·30–0·9	1·2–1·8	0·25	0·10
2036	0·50	0·50	2·2–3·0	0·10–0·40	0·30–0·6	0·25	0·10
BS 3L73	0·8	1·0	4·3	0·8	0·7	0·2	0·3
6061	0·4–0·8	0·7	0·15–0·40	0·15	0·8–1·2	0·25	0·04–0·35
6151	0·6–1·2	1·0	0·35	0·20	0·45–0·8	0·25	0·15–0·35

6

Titanium Adherends

A. MAHOON

British Aerospace, Weybridge, Surrey, UK

1. INTRODUCTION

Titanium was discovered in the form of its oxides in 1791 but the metal was not isolated from its compounds until 1895 by H. Moissan. It is the ninth most abundant element and fourth most abundant metal on earth.[1] The metal and its alloys have outstanding resistance to a wide range of aggressive corroding liquids and show an excellent combination of high strength-to-weight ratio, good fracture toughness and creep resistance. Titanium is inherently a very reactive metal and requires sophisticated and expensive extraction, melting and fabrication techniques. Because of the high cost, it is mostly used only where no other material can perform adequately. It is a common practice for the development of new engineering materials to be sustained by an industry demanding maximum standards of quality and performance. Those demands force the particular industry to fund the developing costs of that new material and its technology; in the case of titanium that role has been performed by the aerospace industry. In addition to its aerospace applications, nowadays titanium finds wide uses elsewhere, for instance in the chemical, petrochemical, process engineering, nuclear and power generation industries, in defence systems and in the medical field. Its application in the thermal and nuclear power generation industries is on the increase[2] because of its high corrosion and oxidation resistance. Parts of fossil-fuelled steam and nuclear turbines (including rotating blades) are now made out of this metal. In the heat exchanger field, the application of titanium-based materials is expanding into areas such as condenser tubes and evaporators, pure-water manufacturing, cooling water towers, fluid coolers for hydraulic controls, etc. Section 8 of this chapter, on high-temperature adhesives used for bonding titanium, reflects the interest in using titanium at elevated temperatures in aerospace and other applications. Many interesting medical applications have also emerged in recent years, for example titanium is being used in the manufacture of heart-pacer cases, heart valves, compression plates for bone fractures, artificial limbs and for opthalmic instruments.

By far the largest user of titanium is still the aerospace industry. Civil and military aircraft, helicopters, rockets, missiles and jet engines all use titanium in some form or the other. A Boeing 747 aircraft has some four tons of titanium incorporated into its structure and single structural components such as landing gear beam assembly and chord-webs can be as long as 7 m. In the future design of aircraft, because of its compatible thermal properties and electrochemical nature, titanium will be more widely used in the manufacture of hybrid structures based on carbon fibre and other composite materials. Since hybrid structures are mostly manufactured by adhesive bonding, this joining technique for titanium aircraft components is gaining significant importance.

2. JOINING OF TITANIUM

Fabrication of titanium involves heavy equipment and large capital investment because of the special processing procedures required for the metal. Due to the high cost of titanium the fabrication processes are aimed at avoiding metal wastage. For the production of heavy sections like beams, bulk heads, chords, engine vanes, turbine blades, etc., fabrication processes based on special casting techniques, isother-mal forging and hot isostatic pressing of powders are employed.[3] However, in many applications parts made of titanium have to be assembled from thin-gauge materials which do not lend very well to the processes used for the fabrication of heavy sections. The following paragraphs briefly outline some of the techniques suitable for joining of titanium in thin sections.

2.1. Mechanical Fastening

Titanium sheets may be joined by riveting or bolting. Since titanium exhibits a high electropositive potential in neutral aqeuous environ-ments, it induces galvanic corrosion of common rivet materials such as steel and aluminium; therefore rivets and bolts for joining of titanium have to be manufactured from galvanically compatible materials. Joining by riveting requires drilling, which can also present problems because of the galling nature of titanium.

2.2. Welding and Brazing

Titanium is successfully joined by the conventional metal joining techniques based on welding and brazing processes. A variety of fusion welding processes is employed, e.g. tungsten inert gas process, electron beam welding and, more recently, laser welding. Sheets of titanium are also joined by resistance welding in the form of spot or stitch welds. Brazing of titanium is achieved by torch, furnace or induction processes. Because of the highly reactive nature of titanium and its fracture sensitivity to hydrogen and oxygen pick-up, most of the processes based on heating require inert gas atmospheres and very clean conditions.

2.3. Diffusion Bonding

In this process the surfaces to be joined are placed in contact, heated to about 950°C and pressed together either by mechanical means or by an inert gas pressure. Diffusion of metal across the interface produces the bond. The diffusion is aided by the fact that the oxide film, which in the case of other metals limits the interdiffusion process, is absorbed within titanium thereby producing bare metal surfaces in close contact with each other. This phenomenon results in good bonding between the two metal surfaces.

2.4. Superplastic Forming and Diffusion Bonding

Titanium and its alloys become highly plastic when heated to about 950°C in an inert atmosphere and in that state can be shaped into any configuration. In this shaping/joining process normally a heated sheet of metal is subjected to high gas pressure in an inert atmosphere. The gas forces the sheet to take the shape of the heated die. Since both superplastic forming and diffusion bonding occur at the same temperature, the two processes have been combined to produce a range of shapes in thin sections. The technology is still under development and has much potential for high-temperature applications where common adhesive systems are found to be inadequate.

2.5. Weld Bonding

This joining process is based on the combination of spot welding and adhesive bonding. The components to be joined are pretreated in a specific manner, layered with the adhesive, spot welded at preselected areas and then heated under pressure to cure the adhesive. The technique is believed to offer a reliable solution to overcome the problem of service failures which can result either from poor surface preparation at the assembly stage or the loss of adhesion due to environmental degradation at the adhesive/metal interface. The problem of developing a suitable 'compromise' pretreatment for both techniques is one reason this expectation has yet to be realised.

2.6. Adhesive Bonding

Joining of titanium with adhesives has primarily been developed for aerospace applications. However, because of the extending use of adhesive bonding to other technologies, the technique holds promise for application in the chemical, petrochemical and power engineering industries, e.g. lining of thick-walled steel reaction vessels with chemical-resistant thin titanium foils. In the past adhesive bonding has primarily been used where no other joining technique was found to work but, with the development of special adhesives and of the technology at large, adhesive bonding is gaining recognition as a joining technique in its own right. As discussed in detail in Chapter 8, in aerospace applications adhesive bonding offers many advantages when compared with the conventional joining techniques of riveting or bolting:

(a) Improved compression strength due to the absence of inter-rivet buckling and improved stiffness because of the initial high buckling stresses.

(b) Better tolerance to large flaws and safety enhancement of the structure due to a change in the mode of failure, i.e. a slow rate of crack propagation.

(c) Higher fatigue resistance because of the absence of rivet holes and other stress raisers.

(d) Weight saving and the elimination of galvanic corrosion in the absence of rivets.

(e) Smooth contours and continuous external surfaces providing improved aerodynamic characteristics.

(f) Even distribution and better dissipation of localised stresses.

(g) The use of thin-gauge material without any detrimental effect on the structural stiffness and strength.

However, adhesive bonding does require better coordination between design, tooling and manufacturing relative to other methods of fastening. Another limitation of adhesive bonding is that the process often requires large autoclaves or presses, vacuum generation systems, clean controlled atmospheres and special tooling.

3. DURABILITY OF TITANIUM ADHESIVE-BONDED JOINTS

Experience has shown that an initial high joint strength does not necessarily ensure that the joint will be able to maintain its integrity over a long period of time when subject to high stresses in a hot/humid environment (see Chapter 1, Fig. 4). Now, as discussed in earlier Chapters, the main site of environmental attack in metallic bonded substrates is usually at, or close to, the adhesive (or primer)/oxide interface and this is the case in adhesively-bonded titanium joints. Since the interfacial regions are of prime interest, in order to improve the durability of titanium joints recent efforts have concentrated on the development of improved surface preparations. Surface treatment influences the following physico-chemical characteristics of the surface: (a) surface wettability; (b) macro- and micro-roughness; (c) chemical composition; (d) crystallographic nature; and (f) stability of the oxide.

Considering these characteristics in turn, then wetting of the titanium surface by the adhesive is important because it controls the extent of adhesive/adherend interfacial contact. The extent of wettability and the rate of wetting is affected by the roughness, the shape of channels, the size of channels and the capillarity of the titanium oxide.[4] Roughness of the surface may result in mechanical interlocking of the adhesive to the adherend, or at least to an increased area of contact between the two surfaces. Morphological studies of the surfaces prepared by a number of pretreatments have suggested that one of the most important factors responsible for providing durability to the titanium adhesive bonded joints is the micro-roughness of the surface.[5] However, this particular approach does not take into account the important role played by the chemical and crystallographic nature of the oxide. Titanium, when treated in a range of different chemicals and by varying processing techniques, produces surfaces with widely differing chemical compositions and oxide stoichiometries.[6] Since the nature of chemical bonds can be influenced by these morphological features, a relationship must exist between the chemical composition of the surface and the joint durability. By preparing surfaces of titanium in a range of solution concentrations derived from the same basic constituents it has been shown that, in addition to surface roughness, chemical composition of the oxide has a strong influence on the bond durability.[7] In another study[8] variations in the crystalline structure of the titanium oxide have been found to influence the permanence of adhesion and the joint durability has been prolonged by stabilising a certain crystallographic modification of the oxide.[9] Scanning electron microscopy studies of the surfaces prepared by different processes have shown that the oxide is unaffected by exposure to humidity and it has been suggested[10] that the inherently high stability of the oxide to moisture degradation is responsible for the good durability that can be attained when adhesively bonding titanium.

4.   HISTORY OF PREBOND TREATMENTS

Adhesive-bonded helicopter blades consisting of titanium skins in the form of sandwich panels have been in use since the 1950s.[11] The earliest prebond treatments were based on cleaning and etching in alkaline mixtures.[8] These processes provided good joint strength but were sensitive to the chemical composition of the titanium alloy and, within the same alloy, were sometimes affected by batch-to-batch variations. Following these difficulties, an etch used for pickling stainless steel and based on mixtures of nitric–hydrofluoric acids was adopted for treating titanium surfaces prior to adhesive bonding.[12] However, when the intermediate curing-temperature adhesives were introduced the acid-mixture treatment presented some difficulties. The surfaces produced were wetted by adhesives but the joint strength with titanium alloys was much lower than that attained with aluminium alloys bonded with the new adhesive systems. The need for a better surface preparation led to the development of the phosphate–fluoride process.[11] This process had been the most widely used surface preparation in the aerospace industry and is covered by ASTM Specification D-2651 and US Military Standard Mil A-9067. In the late 1960s certain United States Army helicopters operating in South-East Asia suffered severe debonding problems in sandwich panels of titanium and glass-reinforced epoxy composite skins bonded to aluminium honeycomb core. The failures were attributed to the ingress of moisture to the interface and joint failure was considered to have been accelerated due to the conjoint effect of moisture and stress. Due to the seriousness of the problem a number of research programmes were initiated to understand the mechanism of failure in hot/humid environments and to improve the environmental resistance of titanium adhesive-bonded joints. One such programme resulted in the development of a modified phosphate–fluoride process.[9] A little later, two commercial surface preparations, namely Pasa Jell[13] and VAST (Vought Abrasive Surface Treatment) processes,[14] became available as alternatives. Because of the expanding use of adhesive bonding in the aerospace industry and the adoption of titanium for the construction of hybrid structures based on carbon-fibre composite materials, new requirements have emerged in the form of higher joint strength and improved environmental resistance. To help fulfil these requirements, many surface preparations have been developed in the last decade,[15–19] as discussed below.

5. CLASSIFICATION OF SURFACE PREPARATIONS

The various surface preparations specifically developed for improving adhesive bond strength and durability can be broadly classified into four techniques depending upon the nature of the process: (a) mechanical, (b) chemical, (c) mechanico-chemical, and (d) electro-chemical. Table 1 lists the processes according to this classification.

5.1. Mechanical Processes
These processes basically employ mechanical abrasion or impact force to clean the surface. Irrespective of whether the mechanical cleaning is conducted in a dry or a wet medium, it results in a macroscopically rough surface containing a thin oxide film.

5.2. Chemical Treatments
Most of the chemical solutions function by etching the existing oxide. If the solutions are reducing in their character, a fresh thin oxide film is produced after the metal is removed from the solution, whereas in the presence of an oxidising agent, e.g. nitric acid in hydrofluoric acid or hydrogen peroxide in caustic soda solution, the etching is accompanied by the formation of a relatively thicker oxide. The newly formed oxide has its own characteristic appearance depending upon the solution

TABLE 1

CLASSIFICATION OF SURFACE PRETREATMENTS FOR ADHESIVE BONDING
OF TITANIUM

Classification	Process	Results	
		Joint strength	*Joint durability*
Mechanical	Alumina blasting	Good	Adequate
	Mechanical abrasion	Poor	Poor
Chemical	Alkaline cleaners	Poor	Poor
	TURCO 5578	Adequate	Adequate
	Nitric–hydrofluoric acid etch	Adequate	Poor
	Phosphate–fluoride	Adequate	Poor
	Modified phosphate–fluoride	Adequate	Slightly better than the conventional
	Pasa Jell 107	Adequate	Adequate
	Alkaline-peroxide etch	Good	Good
	Activated chemical oxidation	Good	Good
Mechanico-chemical	VAST process	Good	Poor
Electrochemical	Anodising in chromic acid–fluoride mixture	Good	Good
	Anodising in alkaline-peroxide solutions	Good	Good
	Cathodic deposition from non-aqueous solutions of aluminium nitrate	Adequate	Adequate

composition. In the case of conversion coating processes, such as phosphate–fluoride treatment, a separate thick oxide layer is produced during the conversion stage of the process.

5.3. Mechanico-chemical Surface Preparations

The VAST process is the most commonly known one which combines mechanical abrasion with chemical etching to produce an adhesive bonding surface. The process is believed to rely on the etching of the surface to produce micro-roughness in addition to macro-roughness resulting from the blasting process. In conventional surface preparations, alumina blasting is sometimes employed prior to the chemical or electrochemical treatment of titanium surfaces to improve bond strength.

5.4. Electrochemical Based Processes

Electrochemical reactions where a metal can be made to act as an anode or a cathode of a cell have been used for cleaning, etching and oxidation of metal surfaces. Anodising solutions (where the metal acts as an anode) must produce a controlled rate of dissolution accompanied by oxide formation. In the chromic acid–fluoride process, the fluoride aids the dissolution whereas the chromic acid contributes to the oxide formation under anodic control. For adhesive bonding a porous and open oxide structure is considered to provide good adhesion (see Part I) and any anodising solution for prebond treatment of titanium must balance the dissolution and oxidation reaction so that a porous oxide results.

Cathodic cleaning, as such, has rarely been employed for the surface preparation of titanium prior to adhesive bonding; however, cathodic deposition of a metal oxide has been shown to be a good prebond treatment. The surface preparation is based on depositing aluminium oxide from non-aqueous solutions of aluminium salts on to the titanium adherends.

6. CURRENT INDUSTRIAL SURFACE TREATMENTS

6.1. Mechanical Cleaning

Abrasion of titanium surfaces serves two functions, i.e. cleaning and roughening. Cleanliness, in addition to surface roughness, has always been considered to be one of the most important factors in adhesive bonding. Roughening of the surface by hand using a variety of abrasives has been an established technique for *in-situ* repairs of adhesive bonded parts. Wire brushes, sandpaper and proprietary scouring materials like Scotch-Brite (3M Company) are commonly used. A better control and efficiency in the cleaning process is achieved by blasting the surface with the abrasive media using either pressurised air or water. The process is also sometimes called honing. The commonly used abrasive media are alumina powder (different sizes), fine pumice stones and nutshells. It has always been considered that, although abrasive treatment improves the initial dry strength, the environmental resistance of surfaces prepared by this means is poor. This may be true for aluminium adhesive bonding surfaces (see Chapter 5), but evidence exists[15] that in the case of titanium a combination of mechanical cleaning and adhesive selection results in adequate joint durabilities

TABLE 2
JOINT DURABILITY OF ALUMINA BLASTED 6Al-4V-Ti ALLOY BONDED WITH
BSL 312/5 ADHESIVE[a] AND EXPOSED TO 95–100% R.H. AT 50°C[15]

Exposure time (h)	Lap shear strength (MPa)	Peel strength (N/2·54 cm width)
0 (initial strength)	54·7	130
500	42·5	116
1000	36·9	116
4000	22·9	67

[a] Ciba–Geigy; modified epoxy film.

being attained. Table 2 shows the joint durability of 6Al-4V-Ti alloy after alumina blasting (180–200 mesh size) at 60–80 psi for 5 min and then bonding with an epoxy adhesive.

6.2. TURCO 5578 Etch

Degreasing and cleaning of titanium in alkaline solutions such as caustic soda, sodium metasilicate and pyrophosphate mixtures, is an essential step in the surface preparation prior to adhesive bonding. Moderate alkaline cleaning on its own produces neither adequate bond strength nor environmental resistance.[12,20] However, strong alkaline solutions which cause etching of the surface have been shown to produce durable bonds.[5] One such proprietary mixture is based on TURCO 5578, a product of TURCO Products Inc., USA. Aqueous solutions of 6–8 (wt/vol)% concentration maintained at 80–95°C have been recommended. After 10 min in the etchant ca 51 ± 2 ppm hydrogen is picked up by titanium[21] which can cause embrittlement and delayed cracking of the metal.

6.3. Acid Etchants

A number of acid mixtures based on nitric–hydrofluoric acid,[12] hydrochloric–orthophosphoric acid,[13,22] hot sulphuric acid[11] and sodium dichromate–chromic acid mixture,[23] have been tried as pre-bond treatments for titanium. Most of these surface preparations provide adequate dry joint strength but the durability is very poor. An additional problem associated with these surface preparations is the pick-up of hydrogen. Acid etchants on their own have not been recommended for surface preparation but they do form an important intermediary stage of other treatments. The mixture commonly used for these applications consists of 3% hydrofluoric acid and 15% nitric acid. The mixture results in the minimum hydrogen pick-up.[24,25]

6.4. Phosphate–Fluoride Process

Chemical conversion coatings for adhesion of organic resins to titanium surfaces have evolved from the application of these coatings to improve lubrication and reduce galling of titanium during metal-to-metal abrasion. The mixtures for film formation are based on combinations of trisodium phosphate, disodium tetraborate, potassium fluoride

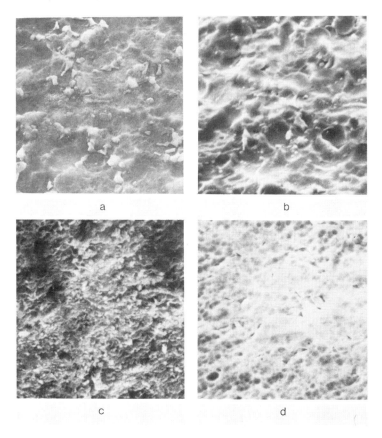

Fig. 1. Scanning electron micrographs of 6Al-4V-Ti alloy surface subjected to various surface preparations (magnification ×2500). (a) Untreated surface; (b) nitric acid–hydrofluoric acid etch; (c) phosphate–fluoride process; (d) chromic acid–fluoride anodise.

and hydrofluoric acid.[26] For adhesive bonding applications, an aqueous mixture containing 5% trisodium phosphate, 2% potassium fluoride and 2·6% hydrofluoric acid is employed.[11] A process based on pre-etching in 3% hydrofluoric acid and 15% nitric acid followed by a 2 min dip in the conversion coating solution described above has become known as the phosphate–fluoride process (US Patent 2 864 732). The rinsing procedure subsequent to the conversion stage influences bond strength[11] and a 15 min soak in deionised water at 60°C has been recommended. The conversion coating is basically an oxide of titanium having anatase crystallographic structure. Thickness of the coating ranges from 1500 Å[27] to 2950 Å[28] and contains Ti, O, P, F, Al and C. Topography of the surface is shown in Fig. 1.

6.5. Modified Phosphate–Fluoride Treatment

Anatase oxide produced by the phosphate–fluoride process slowly reverts to the rutile form on exposure to warm/moist environments.[28] This conversion is accompanied by a decrease in volume of about 8% which induces stresses at the adhesive/oxide interface thereby accelerating the joint failure. It is believed that, if the anatase structure can be stabilised, the subsequent joint durability of the phosphate–fluoride-treated surface is enhanced. This has been achieved by adding 0·75% sodium sulphate into the conversion coating mixture.[9] Table 3 compares the joint strength and environmental resistance of 6Al-4V-Ti alloy after the conventional and the modified phosphate–fluoride treatment. Because of the development of better surface preparations, the phosphate–fluoride-based treatments are slowly becoming obsolete.

TABLE 3

JOINT DURABILITY OF 6Al-4V-Ti ALLOY TREATED BY THE CONVENTIONAL AND THE MODIFIED PHOSPHATE–FLUORIDE PROCESS AND EXPOSED TO 95–100% R.H. AT 50°C.[15] (ADHESIVE, BSL 312/5)

Exposure time (h)	Lap shear strength (MPa)		Peel strength (N/2·54 cm width)	
0 (initial strength)	49·7	53·1	138	143
500	41·1	41·6	18	67
1000	28·5	34·8	18	13
4000	15·4	19·0	—	—

6.6. Pasa-Jell Treatment

Pasa-Jells are proprietary chemicals marketed by Semco Sales and Services Co., Los Angeles, USA. Pasa-Jell 107 is the recommended product for prebond treatment of titanium. The formulation is available either as a thixotropic paste suitable for brush application or as an immersion solution for tank treatment. Approximate chemical constituents of Pasa-Jell 107 are 40% nitric acid, 10% combined fluorides, 10% chromic acid, 1% couplers, and balance water. The immersion process needs non-metallic tanks made of PVC, polyethylene or polypropylene. A recommended mixture uses 1 : 1 dilution for 12 min. With the use of thixotropic paste, a reaction time of 10–15 min is claimed to give durable bonds. Pasa-Jell when used in combination with pre-etching in TURCO 5578 etch is referred to as MCAIR (McDonnell Aircraft Corporation) process for adhesive bonding.[29] The surface treatment compares well with other surface preparations.[5,27,29] The process produces an amorphous looking oxide containing O, Ti, N, Si, C, F and Cr. The oxide has anatase structure which is stable up to 175°C and converts to rutile at 350°C.[14]

6.7. VAST Process

VAST stands for Vought Abrasive Surface Treatment and is a development of Vought Systems of LTV Aerospace Corporation, Dallas, USA.[14] In this process the metal is blasted in a specially designed chamber with a slurry of fine abrasive containing fluorosilicic acid under high pressure (US Patent 3 891 456). The alumina particles are about 280 mesh in size and the acid concentration is maintained at 2%. The process produces a grey smut on the surface of 6Al-4V-Ti alloy and a post-treatment rinse in 5% nitric acid is required to remove the smut. Joint strength and durability of the VAST-treated surface is considered to be slightly lower than that after the TURCO 5578 etch but better than that from the phosphate fluoride process.[17] The film produced is crystalline in nature having an anatase structure containing Ti, O, Si, F, Pb and C. The oxide is stable up to 175°C but starts converting to rutile at higher temperatures.[14] Because of its special equipment requirements the process has found limited application.

6.8. Alkaline-Peroxide Etch

When titanium is immersed in alkaline hydrogen peroxide solutions, depending upon the concentration of sodium hydroxide and hydrogen

TABLE 4

JOINT DURABILITY OF 6Al-4V-Ti ALLOY TREATED IN ALKALINE-PEROXIDE
ETCH, BONDED WITH BSL 312/5 ADHESIVE AND EXPOSED TO
95–100% R.H. AT 50°C.[15,19]

Exposure times (h)	Lap shear strength (MPa)	Peel strength (N/2·54 cm width)
0 (initial strength)	49·4	140
500	42·0	138
1000	41·4	138
4000	33·9	100

peroxide, the metal is either etched or oxidised.[30] Those concentrations which produce grey oxides have been found to produce adhesive wettable surfaces.[31] The metal has to be treated for 10–36 h at ambient temperatures to produce high-strength and durable joints; the use of these mixtures for industrial surface preparations is therefore limited. However, good bonding surfaces are produced within 20 min if the temperature is raised to 50–70°C.[15] This procedure has commonly become known as the RAE etch[7,15] and has been developed by the conjoint efforts of British Aerospace and Royal Aircraft Establishment. A recommended mixture is based on 2% caustic soda and 2·2% hydrogen peroxide (30 volume). Joint strength and environmental resistance are shown in Tables 4 and 5. Morphological features of the

TABLE 5

JOINT DURABILITY (% RETENTION) OF 6Al-4V-Ti ALLOY TREATED IN ALKALINE-PEROXIDE ETCH, BONDED WITH BSL 312/5 ADHESIVE AND EXPOSED TO NATURAL WEATHERING CONDITIONS (RAE, FARNBOROUGH, ENGLAND; INNISFAIL AND CLONCURRY, AUSTRALIA[32])

Applied stress level during exposure (% of initial strength)	Exposure (years)											
	0·5	1	2	4	0·5	1	2	4	0·5	1	2	4
	RAE (temperate)				Innisfail (warm/wet)				Cloncurry (warm dry)			
0	93	94	97	93	97	94	93	92	101	99	94	95
5	97	98	92	90	97	94	95	92	102	99	96	94
10	98	99	92	91	98	95	93	91	99	100	96	95
25	96	98	96	90	98	97	95	91	100	100	97	94

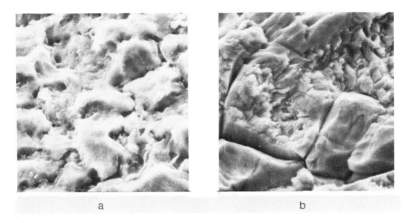

a   b

Fig. 2. Scanning electron micrograph of the alkaline-peroxide etched surfaces (magnification ×2500): (a) 6Al-4V-Ti alloy; (b) commercial purity titanium.

surface are dependent upon the chemical composition of the metal and the treatment mixture. As shown in Fig. 2 on the 6Al-4V-Ti alloy, a continuous oxide about 2000 Å thick is produced whereas titanium of commercial purity gives an etched surface possessing a 1700 Å-thick oxide.[32] The oxide is amorphous in nature and contains 45% O, 22% Ti, 21% Ca, 10% C, <5% Na and S and <2% Cl.

High bond strengths (50–55 MPa) are produced if the titanium surface is subjected to alumina blasting prior to treatment in the alkaline-peroxide mixture. The process offers many advantages over the acid-based treatments for example, the chemical constituents are less toxic, the treatment does not require acid-resistant containers, the process is free from hydrogen pick-up and the waste disposal is simple. The process, however, is limited to batch production because of the high instability of hydrogen peroxide at the elevated temperature of the operation.

6.9. Chromic Acid–Fluoride Anodising

This surface preparation can be regarded as the first-generation anodising procedure providing joint strength and durability equivalent to the alkaline-peroxide etch but much better than the phosphate–fluoride treatment. The process is a development of the Boeing Company, Seattle, Washington, USA and is covered by US Patent 3 959 091. The patent claims that a porous adhesion-promoting oxide is

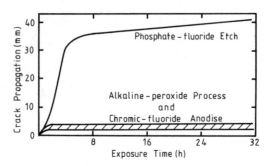

Fig. 3. Influence of surface treatment on the rate of crack propagation in wedge specimens (see Chapter 1) exposed to 95–100% R.H. at 50°C (initial crack length = 25 mm).

produced on titanium and its alloys when anodised from 5–40 V in 5% chromic acid containing fluoride (from addition of hydrofluoric acid) at a level such that the anodising current is maintained at 3·0–600 mA/cm². Detailed investigations to optimise the anodising conditions[18] have revealed that the process produces a brittle oxide when the adherend is anodised at voltages greater than 10 V. It has been suggested that the best results are produced between 3 and 5 V within a 20 min anodising time. Joint durability of 6Al-4V-Ti alloy anodised at 5 V is shown in Fig. 3 relative to the phosphate–fluoride treatment and the alkaline-peroxide etch. No information on the crystallographic nature of the oxide is available; however, the oxide is rough in its appearance,[5] about 2500 Å thick[27] and contains O, Ti, N, C and F.[17]

7. RECENT DEVELOPMENTS IN SURFACE PREPARATIONS

Because of the increased importance of titanium as a structural material in the new generation of civil and military aircraft, more effort has been directed towards the development of improved adhesive bonding technology. Both new high-temperature resistant adhesives[33,34] and better surface preparations need to be developed to meet the challenge of these new requirements.

7.1. Cathodically Deposited Aluminium Oxide

Cathodic deposition of metal oxides from alcoholic solutions containing inorganic nitrates has been developed by Northrop Corporation, Los Angeles, California (US Patent 409475, June 1978). The cathodically deposited oxides are adhesive-wettable and provide good environmental resistance in hot and humid conditions. An evaluation of this process relative to the other surface preparations has shown[35] that joint strengths and durabilities better than the VAST process and the TURCO 5578 etch are produced. A recommended solution is based on 10 g of hydrated aluminium nitrate, $Al(NO_3)_3.9H_2O$, dissolved in 1 litre of isopropyl alcohol. The oxide deposition is achieved by making the metal a cathode of an electrolytic cell maintained at 30 V.

7.2. Activated Chemical Oxidation

Although joint strength and durability attained by treating titanium in the alkaline-peroxide etch is satisfactory, the process requires long heat-up times, has a high hydrogen peroxide consumption rate and presents difficulties when incorporated into a continuous adhesive-bonding assembly line. This has necessitated the development of a room-temperature operated process.

Studies on reaction kinetics of the alkaline-peroxide etch have shown that during the 20 min period the concentration of hydrogen peroxide decreases from 2·5 to 0·1%. The surface of the adherend is etched at the initial stages and oxidised near the end of the processing time.[36] These investigations have indicated that the oxide formation is related to the rate of hydrogen peroxide decomposition and, for producing the oxide at ambient temperatures, the rate of hydrogen peroxide decomposition has to be artificially increased. This is achieved by adding heavy metal ions to relatively stronger solutions of hydrogen peroxide (5–10%). A range of sodium hydroxide and hydrogen peroxide concentrations provide reasonable joint durability when 55 mg/litre of manganese is employed as an activator (Fig. 4).

Sometimes adhesive bonding involves very small parts in large numbers, e.g. washers, ferrules, skins, etc. Jigging of such small parts is time-consuming and dismantling after the surface treatments may damage the treated surface. In order to overcome these production problems a process based on a barrel operation has been developed.[37] The parts to be treated are placed in a perforated barrel and lowered into a tank containing the treatment mixture. The barrel is inclined at

A. Mahoon

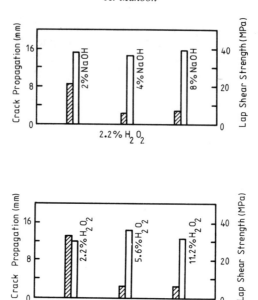

Fig. 4. *Joint strength and durability of 6Al-4V-Ti alloy prepared by activated chemical oxidation in 55 mg/litre $MnSO_4$ (Boeing wedge test using an exposure to 95–100% R.H. at 50°C for 5 h). Hatched bars, crack propagation; empty bars, initial lap shear strength.*

an angle of 45° and rotated at 15–30 cpm. A treatment time of 1 h in a mixture recommended for the treatment of larger articles has been found to produce durable adhesive joints.

7.3. Anodising in Alkaline-Peroxide Solutions

Anodising in various acids[20] and alkaline solutions[38] has been investigated by many researchers as a pretreatment for adhesive bonding of titanium. Most anodising processes produce adhesive-wettable surfaces but the subsequent joint durability is questionable. Anodising is an attractive process as a prebonding treatment because it eliminates the absorption of hydrogen. Anodising in mixtures of sodium hydroxide and hydrogen peroxide produces adhesive-wettable surfaces but only a few concentration combinations provide durable adhesive joints.[16] The effect of solution concentration on joint strength and durability is shown in Fig. 5. Anodising times range from 20 to 40 min and voltages between 5 and 15 V provide durable bonding surfaces. Depending

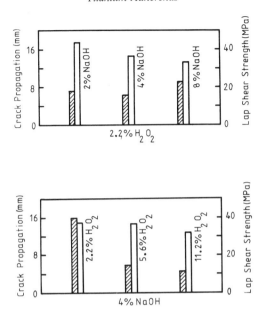

Fig. 5. Joint strength and durability of 6Al-4V-Ti alloy anodised at 5 V for 30 min (Boeing wedge test using an exposure to 95–100% R.H. at 50°C for 5 h). Hatched bars, crack propagation; empty bars, initial lap shear strength.

upon the concentration of the constituents, anodising in alkaline-peroxide solutions produces a range of surfaces with varying roughnesses and oxide morphologies. Topographies of the surfaces are shown in Fig. 6, whereas Table 6 relates the oxide chemical composition with joint strength and durability. The oxide produced is amorphous in nature and shows surface features with no similarities to those exhibited by the chromic acid–fluoride anodised titanium.[5]

7.4. Primers for Titanium
The non-corroding nature of titanium means that, unlike aluminium, no special corrosion inhibiting primers are required, even in the presence of a salt-spray environment for example. However, the incorporation of a bonding, or coupling, agent at an adhesive/metal interface may improve interfacial stability and hence the joint durability, as discussed in Chapters 1 and 3. The coupling agent may be used as a primer, or in some instances incorporated into the adhesive when it must obviously diffuse to the interface before it can be effective.

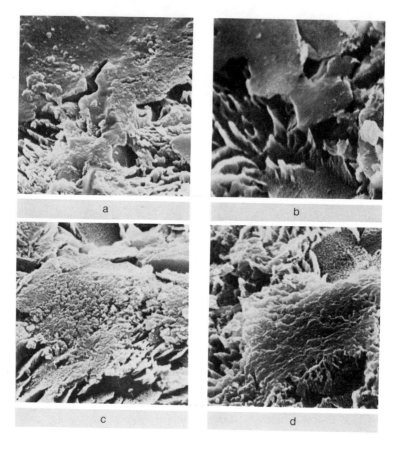

Fig. 6. Scanning electron micrographs of 6Al-4V-Ti alloys anodised in mixtures of sodium hydroxide and hydrogen peroxide (magnification × 10 000). Compositions and voltages were: (a) 2.2% H_2O_2 + 4% NaOH (5 V); (b) 5.6% H_2O_2 + 8% NaOH (5 V); (c) 5.6% H_2O_2 + 4% NaOH (5 V); (d) 5.6% H_2O_2 + 4% NaOH (10 V).

Organo-functional silicones are hybrids of silica and of organic materials related to resins. It is generally considered that they improve the adhesion of metal oxides to resins by way of coupling in the form of metal-oxide silica primary bonds. These compounds are also referred to as coupling agents. Although any polar functional group in a

TABLE 6

EFFECT OF SOLUTION COMPOSITION (ANODISING AT 5 V) ON OXIDE CHEMICAL COMPOSITION, JOINT STRENGTH AND DURABILITY[7]

Solution composition	Elemental concentration (atomic %)				Wedge[a] test (mm)	Initial shear strength (MPa)
	O	Ti	Ca	S		
2·2% H_2O_2 + 4% NaOH	6·8	3·2	0·7	0·5	14	38
11·2% H_2O_2 + 4% NaOH	16·5	9·2	1·8	0·5	5	32
11·2% H_2O_2 + 8% NaOH	9·2	5·2	1·0	0·7	10	33

[a] Exposed to 95–100% R.H. at 50°C for 5 h.

polymer may contribute to bonding to the metal oxide, only a few compounds have been shown to possess the ability to perform as true adhesion promoters by the mechanism of coupling.[39] To improve the joint durability, the adhesion promoter must produce interfacial bonds which are resistant to water. The two families of organic complexes which have shown promise are methyl acrylate–carbon complexes and organo-functional silicones. At an adhesive/metal interface in the presence of silane coupling agents the adhesion between the silane and the adhesive is thought to be due to organic covalent bonds, whereas between the metal oxide and the silane, the bonding is strictly inorganic in character with strong ionic bonding of the type Ti—O—Si.[40] The resistance of the organic covalent bonds to water hydrolysis may be high and the stability of the interface is then dependent upon the water resistance of the bonds between the coupling agent and the metal oxide. It has been shown[41] that, by using γ-aminopropyltriethoxysilane as the coupling agent, a 25% increase in dry strength and a 50% increase in wet strength can be achieved between 6Al-4V-Ti alloy and epoxy adhesives. The increase in strength and durability is assumed to occur due to hydrolytically stable Ti—O—Si chemical bonds between the adhesive and the titanium substrate. The use of adhesion promoters is valuable in situations where surface preparation by conventional techniques is not feasible, e.g. *in-situ* repair bonding. In addition to this application a poor quality control of the pre-bond treatment can sometimes be compensated by the incorporation of an adhesion promoter in the adhesive.

8. HIGH-TEMPERATURE ADHESIVES FOR TITANIUM BONDING AND THERMAL DURABILITY

The use of titanium adhesive bonded structures for high-temperature applications (200–300°C) has been limited due to the rapid degradation of adhesives at these temperatures. Previously developed polyimide adhesives, which are processed by condensation reactions at high temperatures, lead to the evolution of volatile solvents and water vapours. The release of these volatiles may produce porous material which, when used as an adhesive, gives weak joints and poor oxidation resistance.

Table 7 shows the effect of thermal ageing on the loss of lap shear strength of four commercially available adhesives for high-temperature applications.

Because of the increased application of adhesive bonding technology in supersonic-cruise aerospace vehicles, and the use of titanium as heat shields, it has become important to develop high-temperature adhesives with good oxidation resistance and stability and high joint strengths. A new generation of polyimide adhesives[33,34] with these characteristics has recently been developed. Polyimides with terminal acetylenic group have been found to retain 45–50% of their original strength after 1000 h of thermal ageing at 260°C.[42] In another approach, the introduction of perfluoro-alkylene groups into aromatic polyimides has shown a high degree of strength retention after 5000 h at 300°C.[43] To improve the oxidation resistance at elevated temperatures, many of the formulations are pigmented with fine alumina powder.[42] The only high-temperature adhesive which is not based on the polyimide resin is polyphenylquinoxaline. An adhesive based on

TABLE 7

LOSS IN STRENGTH DURING THERMAL AGEING[46] AT 200°C AND 350°C (6Al-4V-Ti ALLOY, STANDARD LAP JOINTS WITH 2·54 × 1·27 cm OVERLAP)

Adhesive	Manufacturer	Adhesive type	Loss in strength (%)			
			At 200°C		At 350°C	
			1 h	1000 h	1 h	1000 h
Imidite 850	Narmco	Polybenzimidazole	28	5	50	—
FM34	American Cyanamid	Polyimide	15	38	50	—
HT424	American Cyanamid	Epoxy–phenolic (aluminium filled)	40	80	90	—
AF131HP	3M Company	Epoxy–Novolac	30	30		—

this hetero-aromatic polymer has been evaluated and found to show a decrease of only 25% in the original strength after 500 h at 370°C.[44] The thermal durability of adhesive bonded joints is also affected by the surface preparation prior to adhesive bonding.[27] Until improved high-temperature stable adhesives and surface preparations are developed, the joining techniques for supersonic cruise vehicles will be limited to superplastic forming/diffusion bonding, welding and brazing.

9. FUTURE DEVELOPMENTS

Adhesive bonding is an expanding technology because of the economic and industrial advantages it offers relative to other joining techniques. In the case of titanium, future developments in surface preparation and adhesive can make a significant contribution to the utilisation of titanium bonded structures.

Prebond surface preparations for titanium, with the exception of one treatment,[7] are normally conducted in toxic acids. Because of the increasing pressure from Government agencies regarding waste disposal and clean working environment, in the future it will become necessary to develop less toxic, and preferably dry, surface preparations. Unlike aluminium alloys where simple mechanical surface cleaning is insufficient for producing durable adhesive joints, titanium can be treated by alumina blasting to produce adequate, but not the ultimate in, strength and durability.[15] In recent years surface preparations based on plasma treatment and ultraviolet–ozone cleaning have gained prominence for adhesive bonding applications in certain industries.[45] The techniques are dry and clean, operate at ambient temperatures and produce least waste. Such techniques offer considerable potential and might be acceptable as surface pretreatments for adhesive bonding of titanium.

In the adhesives field, there is a need to develop better high-temperature resistant resins so that the use of titanium can be further extended to supersonic cruise vehicles and other high-temperature applications. Another increasingly important area in adhesive technology is the need for energy saving. Many structural adhesives have to be cured under pressure at elevated temperatures (120–180°C) to acquire high strength and durability. This particular requirement is time-consuming, costly and poses problems in a production environment. Better room-temperature curing adhesives would not only reduce the cost of manufacture but also eliminate many of the problems associated with the bonding of titanium to other metals and composite materials.

Thus, there is ample scope for further developments in the adhesive bonding of titanium.

ACKNOWLEDGEMENTS

The author is grateful to Mr M. Denney for many helpful discussions and the Board of Directors of British Aerospace for allowing the publication of this text.

Acknowledgements are also due to the Procurement Executive (Ministry of Defence) for financial support for the work on alkaline-peroxide prebond treatments.

REFERENCES

1. Clarke, F. W. and Washington, H. S., *The Composition of the Earth's Crust*, US Geological Survey, Paper 27 (1924).
2. Kimura, H. and Izumir, O., *'Titanium' 80—Science and Technology, Proc. 4th Intern. Conf. on Titanium, Kyoto, Japan*, Metallurgical Soc. of AIME (1980).
3. Eylor, D., Field, M., Frocs, F. H. and Eichelman, G. E., *SAMPE Q.*, **12** (1981), 13.
4. Bateup, B. O., *Int. J. Adhesion Adhesives*, **1** (1981), 233.
5. Ditchek, B. M., Breen, K. R., Sun, T. S. and Venables, J. D., *SAMPE Tech. Conf. Series*, **12** (1980), 882.
6. Wightman, J. P., *SAMPE Q.*, **13** (1981), 1.
7. Mahoon, A., *27th SAMPE Symp. Exhibition* (1982).
8. Wegman, R. F., Hamilton, W. C. and Bodnar, M. J., *SAMPE Tech. Conf. Series*, **4** (1972), 425.
9. Wegman, R. F., Ross, M. C., Slota, S. A. and Duda, E. S., *Picatinny Arsenal Technical Report 4186* (1971).
10. Mahoon, A., *2nd SURFAIR Congress on Surface Treatment in the Aeronautical and Space Industry, France* (1980).
11. Johnson, W. E., Horrigan, J. R., Sumafrank, W. O. and Hooker, J. R., *Fort Worth Quarterly Progress Report on Contract AF33(600) 34392* (1957).
12. Eickner, H. W., *Report 1842*, US Forest Products Laboratory (1954).
13. Walter, R. W., Voss, D. L. and Hochberg, M. S., *SAMPE Tech. Conf. Series*, **9** (1970), 321.
14. Lively, G. W., LTV-Vought Systems Division, *Tech. Report AFML-TR-73-270* (1974).

15. Mahoon, A. and Cotter, J. L., *SAMPE Tech. Conf. Series*, **10** (1978), 425.
16. Mahoon, A. and Kohler, R., British Patent 2074608A (1981); French Patent 81/08123 (1981); German Patent P3116446.3 (1981); Japanese Patent 61515/81 (1981); US Patent 256251 (1981).
17. Wightman, J. P., *SAMPE Q.*, **13** (1981), 1.
18. Locke, M. C., Harriman, K. M. and Arnold, D. B., *SAMPE Symp. Exhibition Series*, **25** (1980), 1.
19. Cotter, J. L., *Joining in Fibre Reinforced Plastics*, IPC Science and Technology Press, London (1978).
20. Snogren, R. C., *Handbook of Surface Preparation*, Palmerton Publishing Co., New York (1974).
21. Layoie, A. D. and Seago, R. A., Bell Helicopter Company (1968).
22. Cagle, C. V., *Handbook of Adhesive Bonding*, McGraw-Hill, New York (1973).
23. Muchnick, S. N., *The Franklin Institute Report 55–87* (1955).
24. Rozenfeld, I. L., Babkin, Yu. A. and Alekseeva, E. I., *Zashchita Metallov.*, **6** (1970), 410.
25. Dass, K. B. and Marceau, J. A., *Corrosion*, **30** (1974), 324.
26. Miller, P. D., Jeffreys, R. A. and Pray, H. A., *Metal Progress* (May 1974), 61.
27. Hendricks, C. L. and Hill, S. G., *SAMPE Q.*, **12** (1981), 32.
28. Hamilton, W. C. and Lyerly, G. A., *Gillette Company Research Institute Report AD724663* (1971).
29. Stifel, P. M., *SAMPE Symp. Exhibition Series*, **19** (1974), 75.
30. Maza, F., *Chimica e Industria*, **43** (1961), 1293.
31. Allen, R. W., Alsalim, H. S. and Wake, W. C., *J. Adhesion*, **6** (1974), 153.
32. Cotter, J. L. and Mahoon, A., *Int. J. Adhesion Adhesives*, **2** (1982), 47.
33. Bilow, N., Landis, A. L. and Miller, J. L., US Patents 4075111 and 4108836 (1980).
34. St Clair, A. K., Slemp, W. S. and St Clair, T. L., *Adhesives Age*, **22** (1979), 35.
35. MacKay, J., *Joint Conf. SAMPE and ASTM* (1980).
36. Mahoon, A., unpublished work.
37. Mahoon, A. and Pullen, T., *British Aerospace Report AL/MAT/3732* (1981).
38. Felson, M. J., *SAMPE Tech. Conf. Series*, **10** (1978), 100.
39. Walker, P. J., *J. Coatings Technology*, **52** (1980), 33.
40. Pleuddmann, E. P., *Int. J. Adhesion Adhesives*, **1** (1981), 305.
41. Schrader, M. E. and Cardamone, J. A., *J. Adhesion*, **9** (1978), 305.
42. Bilow, N., Landis, A. L. and Boschan, R. H., *SAMPE J.*, **18** (1982), 8.
43. Cotter, J. L. and Hockney, M. G. D., *Int. Metallurgical Reviews*, **19** (1974), 103.
44. Hergenrother, P. M. and Levine, H. H., *J. Appl. Polym. Sci.*, **14** (1970), 1037.
45. Mittal, K. L., *Adhesives Age*, **24** (1981), 18.
46. British Aircraft Corporation, *Aircraft Laboratories Report AL/MAT/3107* (September 1970).

7

Steel Adherends

W. Brockmann

Fraunhofer-Institut für Angewandte Materialforschung, Bremen, Federal Republic of Germany

1. INTRODUCTION

The use of organic structural adhesives for the industrial production of steel joints is first referred to in the paper by Preiswerk and Zeerleder.[1] Shortly after the first introduction of phenolic resins for adhesively bonding aluminium alloys by de Bruyne and others[2,3] during the Second World War, he pointed out the usefulness of epoxide resins as structural adhesives for practically all technical materials, amongst them steel. It was further realised that steel joints could also be produced by using, besides epoxide resins, adhesives based on phenolic, polyester and acrylic resins. Two prominent applications are noteworthy: a bonded steel bridge, constructed in 1955 in Germany, which has served over 12 years without failure,[4] and the deep-sea bathysphere of Piccard with adhesively-bonded steel parts.[5] They both demonstrate the efficiency of the then-new joining technique, which today is a common production method in different areas of the steel manufacturing industry. Currently, the automobile industry uses adhesives to join stiffeners in car bodies or to insert cylinder liners in combustion engines. Driving shafts and hubs are bonded in large steel rollers and joints between hardened steel and cast iron have survived in machine tools for many years.

In contrast with the structural adhesives used in aircraft, the development of adhesives for bonding steel has been governed not only by considerations of highest possible shear and peel strengths, and later good durability, but more by the consideration of simple and economic processing properties. Sophisticated chemical surface treatment procedures are too expensive, autoclaves for curing processes under elevated temperatures and pressures mostly not practicable and

281

curing cycles needing hours to complete are not compatible with the conditions of modern mass production.

In principle, iron and its alloys, particularly the different types of steel, are relatively easy to bond by nearly all available structural adhesives, if the surfaces to be bonded are free from contaminants and corrosion products. They have to be clean in a technical sense and the choice of adhesive can then be made from considerations of the strength and long-term behaviour, the processing conditions, the economics or (a factor becoming more important) the toxicity.

A survey of some of the most important properties of the different types of structural adhesives used today is given in Table 1. The data for the curing conditions and the highest service temperatures are average values and in the columns for strength and durability only qualitative comments are given, which are a result of the author's experience in the adhesives field over a period of 15 years. Obviously, the data given in the table can only give a very rough overview. For example, some cold-curing epoxide resins may be cured at temperatures of 5°C as well as at 200°C, when the curing time decreases from approximately 100 h to a few seconds. These very different curing conditions will influence the mechanical properties of the adhesive.

While the table only includes the solvent-free resins, solutions of resins may also be used. Then the typical curing conditions will change again and so will the technical properties. These few points indicate how large is the number of possible variations in each of these resin systems. If it is economical for a special bonding problem, an adhesive can be formulated with optimum properties for ease of application, strength and durability. Normally, high shear strength may be equated in structural steel bonds with values higher than 10 N/mm^2 and in the case of peel strength with values higher than about 5 N/mm. High durability means, for example, that a well produced steel joint can be used for more than five years in the natural climate of Northern Europe without a loss of strength of more than about 20%.

It is noteworthy that, in the group of cold- or room-temperature curing adhesives, increasing importance is being attached to the polyurethanes and acrylate copolymers. This will be especially true if the manufacturers succeed in decreasing the toxicity and improving the long-term properties in steel joints, compared with two-component epoxys with amine hardeners. The increasing restrictions for materials containing hazardous low molar-mass components also hinder the use of the second-generation acrylics, despite their good processing,

TABLE 1

CHARACTERISTIC PROPERTIES OF ADHESIVES FOR STEEL

Basic resin	Curing Conditions			Strength		Max. service temp. (°C)	Durability in steel bonds	Remarks
	Temp. (°C)	Pressure (daN/mm²)	Time (min)	Shear	Peel			
2-Comp. epoxide	20	0	60–480	High	Low–medium	60–80	Low–medium	Some hardeners toxic
2-Comp. polyurethane	20	0	10–200	High	High	60–80	Low	New developments of better durability expected
1-Comp. epoxide	120	0	30–60	High	High	150–200	High	
1-Comp. phenolic	150	5–8	30–60	High	Low	100	High	Toxic
1-Comp. polyimide	180	5–8	60	High	Low	300	High	Stabiliser toxic
1-Comp. cyanoacrylate	20	0	1–10	Medium	Low	80–100	Low	Glueline thickness must be less than 0·1 mm
1-Comp. diacrylic acid ester	20	0	0·1–10	Medium	Low–medium	80–120	Medium	
2-Comp. acrylics (2nd generation)	20	0	5–60	High	Medium–high	80	High	Activator toxic
1-Comp. hot melts (PA, EVA)	150–200	Contact	0·1–2	Low–medium	Medium	100	Medium–high	Cheap, non-toxic, extremely short setting times possible
1-Comp. plastisoles (PVC)	150–200	0	10–60	Low	Medium–high	80–100	High	Good adhesion on untreated and oily steel surfaces
1-Comp. pressure-sensitive adhesives (tapes)	20	Strong contact	0	Low	Extremely low	100–120	High	

Abbreviations: PA, polyacrylate; PVC, poly(vinyl chloride); EVA, ethyl vinyl acetate.

strength and long-term properties. On this particular topic, the hot melts and pressure-sensitive adhesives, which give strengths sufficient for many applications, represent an important advance in the adhesives field.[6,7]

2. SURFACE TREATMENTS FOR STEEL

For the production of reliable steel joints of high strength and sufficient durability using structural adhesives, surface cleaning and pretreatment processes are usually needed. The first step is the removal of soluble contaminants. This can be done in alkaline aqueous solutions at about 60–80°C, followed by rinsing in de-ionised water and drying in warm air. However, in most cases organic solvents are used in this stage of cleaning; some examples are shown in Table 2. Dipping in the solvent is not usually very successful and the simplest form generally employed is brushing. Better degreasing results are obtained by removing grease in an ultrasonic bath. For mass production, however, vapour degreasing systems should be used. In these totally closed systems the parts to be degreased are stored in a cooled zone over boiling solvent. The solvent vapour condenses on the cool parts in the cooled zone and drops back into the bath.[8] Thus only clean, redistilled solvent comes into contact with the parts.

If steel adhesive joints of medium strength are required and durability against humidity is not needed, cleaning procedures as described above may be sufficient as a surface treatment. However, high strength combined with low scatter of the strength values and reliable resistance against environmental attack in steel joints is only attainable by further surface pretreatment steps. In contrast to aluminium and titanium

TABLE 2

ORGANIC SOLVENTS FOR DEGREASING STEEL SURFACES

Solvent	Boiling temperature (°C)	Hazards
Acetone	56	Flammable
Methylene chloride	42	Highly toxic
Perchloroethylene	121	Toxic
Trichloroethylene	87	Toxic
1,1,1-Trichloroethane	74	Toxic

alloys, where the surfaces are usually treated by the chemical methods mentioned in Chapters 5 and 6, etching procedures for the different types of steels, with one exception, are not recommended. Although some chemical treatments are quoted in the literature,[9] the best results are usually obtained by using mechanical roughening techniques like grinding with emery or shotblasting with corundum (Al_2O_3) as the grit material. Currently these appear to be the best treatments for all kinds of steel, as shown in section 3.2. The only chemical treatment which is generally recommended is for stainless steel and involves etching in an aqueous solution containing 10% oxalic acid and 10% sulphuric acid at 80°C for 10 min, rinsing in water and finally removing the black surface layer ('smut') by brushing.

The mechanical roughening processes, especially shotblasting procedures, remove inactive oxide and hydroxide layers by cutting and deformation processes of the base material leading to a fissured surface topography as shown in Fig. 1. The surface reoxidises almost instantly. The first consequence of such a treatment is an increase in the effective surface area for adhesion reactions compared with a geometrical flat surface. Secondly, the freshly-grown oxide layers show higher chemical reactivity, as can be readily demonstrated by adsorption tests.[10] The higher reactivity of the fresh surface oxide does make it susceptible to corrosion by water, but this reactivity also appears to lead to good durability of the adhesive/oxide interface in humid environments.

It is also of interest that, if liquid water is not present, the chemical reactivity of the pretreated surface, and so its properties for creating good adhesion, are stable over several hours or days under normal indoor conditions. This is demonstrated by the test results plotted in Fig. 2. The specimens of mild steel in this test were shotblasted and stored with the surfaces exposed to room temperature and 60% R.H. prior to adhesive application. After times between some minutes and 330 hours, single-overlap shear specimens were bonded with the steel substrates using Araldite AW 106 (two-component epoxide–aminoamide). On these specimens the initial shear strength was measured and this is plotted as a continuous line in Fig. 2. Further, some of the bonded specimens were aged over a month in a humid climate of 30°C and 95% R.H. and then the residual shear strength was determined, plotted as a dashed line. Initial and residual shear strength increase at first with increasing 'open times' of the surfaces up to 24 h and remain at a high level until storage times of 150 h. With longer storage times the initial shear strength decreases whilst the residual

W. Brockmann

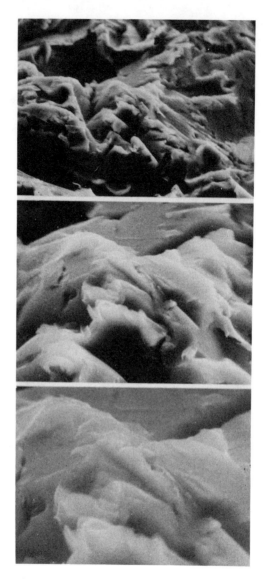

Fig. 1. Scanning electron micrographs of a sandblasted steel surface.

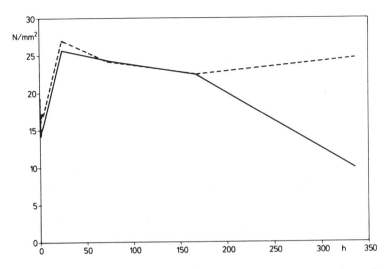

Fig. 2. Shear strength of steel joints (abscissa: exposure time of the surfaces before adhesive application). ———, *Initial shear strength;* – – – –, *residual shear strength after ageing 1 month at 30°C, 95% R.H.*

strength remains nearly constant. An important consequence of these results is that an adhesive need not be applied immediately after mechanical treatment of steel. Similar results were obtained on stainless steel and with ground surfaces.

The significance of the role of the surface treatment will become more evident after the discussion of results in section 3.2. However, it should be mentioned in this context that hot-curing plastisol and epoxide adhesives are currently available which may be used on untreated and oily steel surfaces, for example in the automobile industry.[11] This is discussed in detail in section 3.1.

3. STRENGTH OF STEEL JOINTS

The short-time strength properties obtainable in steel joints, bonded with structural adhesives, are of the same magnitude as in bonded aluminium or titanium joints. Unfortunately, until recently much scientific effort has been devoted to calculating mathematical and theoretical models for the strength of adhesive joints and experimentally determining the deformation properties of the adhesive and adherends.[12-15] On the other hand, examination of adhesion reactions between the adhesives and industrially pretreated surfaces has been largely neglected. The assumption made in most of these calculations is that the adhesive zone is a two-dimensional area between the metal and the primer or adhesive with the presence of one phase not affecting the properties of the other. This is clearly not the real situation. The adhesion zone is actually a three-dimensional zone extending from the base material (sometimes deformed by manufacturing treatments) through oxide layers of different structures (perhaps partly invaded by adhesive molecules) into adhesive layers possessing properties different from the bulk adhesive. This may be illustrated from simple shear tests which readily demonstrate the very significant influence of the boundary reactions. If, for example, single-overlap joints of the same geometry (sheet thickness 1·5 mm, overlap length 10 mm) are bonded using a hot-curing adhesive (Tegofilm, phenolic resin) the shear strength in the case of aluminium (2024 alloy) joints is $39 \, \text{N/mm}^2$ and in the case of mild steel only $30 \, \text{N/mm}^2$. This arises even though the Young's modulus of steel is three times higher than the modulus of aluminium which must theoretically result in a better stress distribution and therefore a higher shear strength for the steel joints. Thus, this observation cannot be explained by only one strength theory. Indeed, the influence of the boundary zone becomes still greater if both types of joints are aged. For example, after 1000 h in a salt-spray environment the residual strength for the aluminium joints was $37 \, \text{N/mm}^2$ and for the steel joints $23 \, \text{N/mm}^2$; in both cases cohesive failure in the adhesive layer occurred.

The above observations also demonstrate that the usefulness of an initial strength value alone for estimating the durability of a bonded joint or for the quality assurance of an adhesive is, to say the least, dubious. The same is true for the long-term loading capacity or fatigue properties if they are measured only under normal, dry conditions. This aspect is considered in detail below.

3.1. Initial Strength

In Figs 3 and 4 the results of bonding mild steel using a hot-cured epoxide resin (based upon a dicyandiamide-cured epoxy resin) and a polyvinylformal-modified phenolic resin respectively are shown. Single-overlap shear strengths were measured as a function of the surface pretreatment employed and shear strength values between 10 and $30 \, \text{N/mm}^2$ were obtained. The dashed line on the top of the

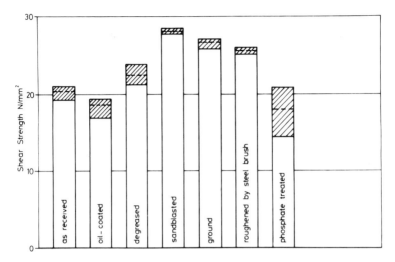

Fig. 3. *Shear strength of single-overlap mild-steel joints. Overlap length 10 mm,
thickness of sheets 0·9 mm. Adhesive, hot-cured one component epoxide resin.*

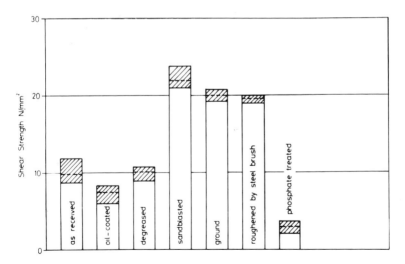

Fig. 4. *Shear strength of single-overlap mild-steel joints. Overlap length 10 mm,
thickness of sheets 0·9 mm. Adhesive, phenolic resin.*

columns marks the average strength value, while the hatched area shows the standard deviation. A comparison of the two Figures clearly reveals the different sensitivity of the two adhesives when bonded against surfaces resulting from the different pretreatments.   For example, the shear strength of joints prepared from phosphate-coated steel is lower than that of mechanically-treated steel, despite the fact that phosphate coatings are very good treatments for painting processes. The low joint shear strengths are due to the relatively low strength of the phosphate coating. Also, the phenolic adhesive is more sensitive to surface pretreatment than the epoxide system, but the former is still used today, for example in the ski-building industry where it is very successful for joining the steel leading-edges to the lower aluminium plates of the construction.

It is interesting to note that, despite the fact that the phenolic resin is a hot-curing system, it shows high sensitivity against contaminants on the surface of a phosphate layer. This statement is also relevant when considering cold-cure epoxide adhesives. Indeed, in the case of oily steel surfaces their use is dubious since the resin cannot easily remove or dissolve the oil during the cold-cure process. Only during a hot-curing process can the adhesive remove the oil and partly dissolve it so that the loss of strength by the oil film is minimised. Finally, the influence of the surface treatments must be mentioned, not only on the absolute shear strength values, but also on their standard deviation. In some cases this is of more importance because a small scatter in strength values means high reproducibility of the bonded system. Again, mechanical roughening of the surfaces by grinding, brushing or sandblasting is the best pretreatment.

Shear strength is the most quoted strength value for adhesive bonded joints. Nevertheless, under a direct tensile load between steel parts, with mechanically roughened surfaces and a bonding area not larger than a few square centimetres, tensile strength values from 10 to *ca* 80 N/mm^2 are obtainable. The tensile strength depends practically only upon the tensile strength of the adhesive itself. The tensile strength of cold-curing resins is in the region of 20 or 30 N/mm^2 and that of hot-curing systems between 50 and 90 N/mm^2. Another strength value of some importance is the peel strength. Some values obtained using the T-peel test with 0·5 mm thick adherends are given in Table 3. Results for four adhesives are shown: a phenolic resin, a one-component epoxide resin, a two-component epoxide resin and a cyanoacrylate. The initial peel strength and the average (propagating)

TABLE 3
T-PEEL STRENGTH OF STEEL JOINTS (SURFACES SANDBLASTED)

Adhesive	Chemical type	Peel strength (N/mm)	
		Initial	Average
Redux 775	Phenolic	4	1·5
Araldite AT I	1-Comp. epoxide	12	10
Araldite AW 106	2-Comp. epoxide	10	2·5
Sicomet 85	Cyanoacrylate	9	1·5

peel strength values, between which differences may occur especially in the T-peel test, are quoted. Of more importance appears to be the average peel strength, from which conclusions concerning the deformation properties of the adhesives are possible. For example, the phenolic system, used for more than 40 years, and the cyanoacrylate are brittle adhesives, whilst one-component epoxide resins exhibit relatively high plastic-deformation behaviour. In comparison modern modified hot-curing epoxide resins (containing a dispersed rubbery phase, Chapter 1) have average peel values in the region of 10 N/mm or more. The difference in peel strength between phenolic resins and modern modified epoxide resins has been one of the reasons why the phenolic systems have been replaced in many applications by epoxide resins, not only in the aircraft industry but also in the ski-building industry.

Considering next static and dynamic fatigue, these properties may be measured on single-overlap steel joints loaded by static or oscillating forces without the influence of hostile environments. If steel bonds are placed under a constant load over 1000 or 5000 h in a dry environment at room temperature, they show in the case of cold-curing adhesives a long-term strength of about 50% of the initial strength, and in the case of hot-curing adhesives between 70 and 80%. In a fatigue test on single-overlap steel joints, with cyclic loads varying cyclically from just above zero to a maximum value, a long-time fatigue strength between 10 and 20% of the initial strength can be expected after 10^6 or 10^7 load cycles.[12]

Initial strength values of bonded joints such as the above, or others obtainable from the adhesive manufacturers or the literature, are useful to compare different adhesives and to demonstrate the effect of such parameters as bondline thickness, overlap length or curing conditions, and, in some cases, also the surface state. All of these may change the observed strength of the bonded joint. However, it is impossible, as shown in section 3.2.2, to describe from initial strength values alone the real efficiency of the adhesive bonding technique as a joining system for steel assemblies if they have to serve over many years at elevated temperatures and especially under humid environmental conditions.

3.2. Long-term Behaviour under Hostile Environmental Conditions

3.2.1. Introduction

It is well known today that, except for the degradation by concentrated acids or bases, X-rays or γ-rays, some organic solvents of high polarity and temperatures of more than 100 or 150°C, the most common and important factor influencing the long-term behaviour of adhesively-bonded metal joints is the presence of humidity or liquid water. The reason for this, apart from the possibility of primary corrosion on the metal parts, is the capability of the water molecules (due to their small dimensions and high polarity) to diffuse into practically all organic materials. This includes the ability to diffuse into crosslinked and/or partly crosslinked organic adhesive layers. Water diffusion into adhesives and also into adhesive layers between metal parts may be measured by gravimetric methods[16] or other techniques using, for example, water-reactive substances mixed into the adhesives.[17] The water diffusion follows, to a first approximation, Fick's law (Chapter 3), so the diffusion velocity and the amount of water invading the polymer are dependent on the concentration gradient of water and the temperature. To give an idea of water diffusion into adhesive layers, a two-component epoxide resin cured with an aminoamide hardener absorbs in an environment of 100% relative humidity and a temperature of 40°C, more than 4% w/w of water. The velocity of diffusion under these conditions in the adhesive layer is in the range of 1 or 2 mm/month. In most cases, the diffusion process ends at a state of equilibrium, the level of which is related to the adhesive and, with only some exceptions, is reversible. Such a diffusion process cannot be avoided by coating the edges of the bonded area by paints or sealings, which only inhibit primary corrosion of the areas and delay slightly the diffusion process.

The free or absorbed water molecules between the adhesive polymer chains can initiate two important mechanisms which change the long-term properties of the bonded joints. The first of these is a slight swelling process leading to a decrease in tensile strength and also, in most cases, to an increase in the plastic deformation properties of the polymer and therefore to higher peel resistance. Chemical reactions between the diffusing water and the organic molecules comprising the bulk adhesive are possible, but as far as is known, do not occur to any great extent and are not very significant in determining the lifetime of bonded joints if crosslinked adhesives are used. The second important mechanism is the capability of water molecules diffusing into the interfacial boundary zone to destroy the adhesion between the adhesive and the metal surface, attack the oxide or degrade the boundary layer of adhesive, which may all lead to the well-known interfacial failures looking from a macroscopic viewpoint like a total separation of the adhesive from the metal surface without changing the nature of the metal surface, as discussed in Chapters 1 and 3. The degradation process in the boundary layer, leading finally to the failure, is not reversible and consequently is the most dangerous mechanism. In this boundary zone the polymer may be affected by the water due to hydrolytic degradation. Such a mechanism will occur more readily if the polymer in these regions has not fully crosslinked, as discussed in section 3.3. Except for primary corrosion on the edges of the bonded joints, electrochemical corrosion processes in the boundary layer between the adhesive and the metal surface cannot occur as long as a solid/solid interface exists, because electrochemical corrosion needs liquid as an electrolyte and this is not present in the undestroyed state of the bondline.

3.2.2. Test Techniques

The effect of water diffusion into the adhesive can be clearly seen from the strength behaviour of bonded joints in humid environments. One of the best laboratory techniques to evaluate these ageing mechanisms is the measurement of the creep of mechanically-loaded shear specimens in an elevated temperature and high humidity environment by optical or electromechanical devices.[18,19] If in these measurements single-overlap joints in the form of standard specimens are used, measurement times of about 1000 h in a constant climate of 30–50°C and 95% relative humidity are usually sufficient. From such tests cold-curing adhesives like aminoamide cured epoxide resins and also second-generation acrylate adhesives showed a long-term strength of only 1 N/mm^2, which is no more than 5 or 10% of the initial shear strength. During the loading process in the humid environment, shear strains in the glue- or bond-line of tan $\gamma = 0.4$ occurred, and under these small loads the creep process came to an end after 100–400 h. Under higher loads, higher creep speeds were observed, and the creep processes did not cease. Practically all specimens under higher loads were destroyed under loading times of 200–300 h. If the temperature of the environment was higher than 40°C, practically no long-term load capability was measurable.

Totally different properties are observed if hot-curing epoxide adhesives, used today for example in the aircraft industry (e.g. rubber-modified epoxide systems), are tested over 1000 h in an environment of 50°C and 95% R.H. These adhesives are able to withstand shear stresses of 15 N/mm^2 and more. This is equivalent to 50 or 60% of the initial strength. The creep processes during this ageing procedure came to an end after about 500 h, and the measured creep strains were in the range of tan $\gamma = 0.5$–1.0.

Only these more realistic test methods, which demonstrate the large differences in the load capability and durability of cold- and hot-curing adhesives, give clear indications of the actual load capability of a bonded metal joint. If an adhesively-bonded steel joint, which has to serve within the temperature range of 30–50°C over long times in the presence of humidity, is designed so that for cold-curing adhesives the load is no higher than 5% of the initial strength and in the case of hot-curing adhesives no higher than 30–50% of the initial strength, then ageing problems and failures within the adhesive over long times should not be observed.

However, this test technique with ageing times of 500–700 h in an artificial climate without condensation of water on the adherends still only gives a broad indication as to the influence of the weakening of the adhesively-bonded joint in the interfacial regions between the adhesive and the metal surface. The resistance of an adhesively-bonded steel joint in this boundary zone is influenced mainly by the type of steel, its surface treatment before adhesive application, and the type of adhesive used in the joint. Further, exposure parameters such as the environment, temperature, level of humidity, etc., are important factors for determining the long-term behaviour. It is usual to investigate the ageing properties of the boundary layer in steel bonds using single-overlap joints and measuring the initial strength and the residual strength after different ageing procedures. It has been suggested that additional mechanical loading of the specimens is not generally needed, since no accelerating effect upon the weakening mechanisms has been found to exist, assuming the loading does not exceed a certain low but realistic level (see above). However, an exception to this may be indicated from the results of long-time tests carried out with unloaded and loaded specimens over a period of six years in Surinam.[20] In this investigation it was found that the residual strength of mechanically loaded specimens decreased after ageing times of some years more rapidly than that of unloaded joints in the tropical climate. The results, however, are difficult to interpret, because the failure mode of the different specimens is not exactly described and corrosion may have played an important role.

In the author's investigations, from which some results are shown in Figs 5–8, with ageing times of one year for the steel joints in an artificial climate (chang. cl.) and a natural climate (nat. cl.) in North Germany, no significant detrimental effect of a cyclic load on the residual strength after 6- or 12-month exposures, compared with the residual strength of unloaded specimens, was detected (see Figs 5 and 7). Examined in these investigations were single-overlap joints produced from mild steel with shotblasted and degreased-only surfaces. The adhesives were a phenolic resin and an epoxide–nitrile system. The strength reduction on unloaded specimens in the case of the phenolic resin was very low on shotblasted steel, whilst in the case of the epoxide–nitrile resin the strength decreased about 30% after half a year of ageing in the natural climate. Compared with the unloaded specimens, the strength reduction in the cyclic-loaded joints was slightly smaller. It should be noted, however, that the cyclic load compared with the initial strength was low: the applied shear stress was

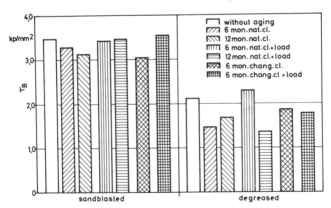

Fig. 5. Initial and residual shear strength of steel joints. Adhesive, phenolic resin.
$(1 \ kp/mm^2 \equiv ca. \ 10 \ MPa.)$

Fig. 6. Fracture surfaces of steel bonds after ageing one year in a natural
climate. Adhesive, Tegofilm (phenolic resin).

alternated, with a frequency of 1 Hz, between 0 and 0·7 N/mm². The
increased level of the residual strength in the loaded specimens may be
due to an increased plasticising effect by invading water in the low-
strained adhesive layer.

3.2.3. Effect of Surface Pretreatment

The results of these tests, shown in Figs 6 and 8, also demonstrate the
influence of the surface treatment on the stability of the boundary

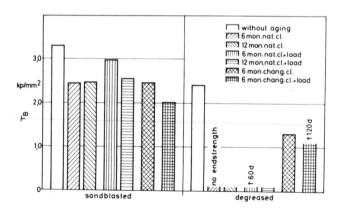

Fig. 7. Initial and residual shear strength of steel joints. Adhesive, epoxide–
nitrile resin. (1 kp/mm² ≡ ca. 10 MPa).

sandblasted

degreased

Fig. 8. Fracture surfaces of steel joints after ageing one year in a natural climate. Adhesive, FM 123/5 (epoxide–nitrile resin).

zone. In the case of the phenolic adhesive in combination with shotblasted steel adherends, after one year of ageing in the natural climate in North Germany and an artificial climate (30°C; 95% R.H.) no interfacial failure occurred in the joints, as may be seen in Fig. 6. In joints with degreased-only steel produced by using the phenolic resin, after one year the whole joint failed by so-called interfacial failure (Fig. 6), and corrosion of the metal surface had started from the edge of the bond. More sensitive to differences in the surface treatment was the epoxide–nitrile resin, originally developed for bonding chemically-treated aluminium alloys. In the case of shotblasting the steel surfaces, small areas of interfacial failure and corrosion initiating from the edges are visible. In the case of degreased-only steel surfaces after 30 and 60 days in the unloaded and loaded state, total delamination and subsequent corrosion of the metal surfaces had occurred (Fig. 8). No connection between the cyclic loading of the specimens during the ageing processes and the occurrence of interfacial failure was observed. These results show that, as discussed in detail in Chapter 4, the type of adhesive, as well as the state of the metal surface, are important parameters for the durability of a bonded steel joint. For both adhesives, a mechanical roughening process of the metal surface by a shotblasting process appears to have a positive effect on the durability of the joints.

The influence of a particular surface treatment on the ageing behaviour of bonded steel joints is not equivalent with all different

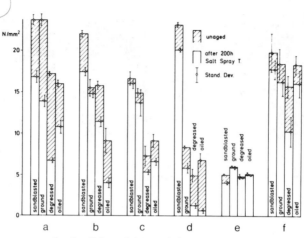

Fig. 9. *Shear strength of mild-steel joints.* (a) *Two-part cold-cured epoxy;* (b) *two-part acrylic;* (c) *anaerobic acrylic;* (d) *cyanoacrylate;* (e) *PVC plastisol;* (f) *one part hot-cured epoxy.*

adhesives, as is shown in Figs 9 and 10. Plotted in these diagrams are the initial and the residual strength values obtained with joints of mild steel and stainless steel employing different adhesives and different surface treatments of the steel adherends. It may be seen that on different steel alloys, different adhesives show a very complex ageing

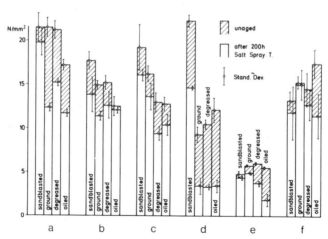

Fig. 10. *Shear strength of CrNi 18/8 joints. For details of parts* (a)–(f) *see caption to Fig. 9.*

behaviour. The degree of ageing is dependent on the state of the surfaces, which were either shotblasted, ground by emery paper, degreased in acetone, or coated with a thin film of corrosion-inhibiting oil. The ageing process in this case was a salt-spray test (ASTM B 117) with a continuous spraying of 5% NaCl solution in 95% R.H. at 35°C. From these results it may be concluded that the durability of steel bonds produced with a hot-curing epoxide resin is better than that of specimens produced with cold-curing epoxide resins and other systems. As expected, the cyanoacrylate adhesive is characterised by relatively low durability, whilst the plastisol adhesive shows very good durability on untreated or oily mild-steel surfaces. These particular properties of the plastisol adhesives are the reason for their wide use in the car-building industry. The relatively low shear strength of the system is, in practice, no disadvantage.

3.2.4. Service Life Prediction

The question that arises now is: how far are such results representative of the real durability of steel joints under other environmental conditions, as well as under longer ageing times?

The rate of degradation in steel joints exposed to humid environments is dependent on the temperature and the water concentration in the glue- or bond-line caused by the environment. Figure 11 shows a comparison of strength values in unaged and aged states after exposure in the natural climate in North Germany (c) and in an artificial climate at 40°C and 95% R.H. (a). The environmental temperatures during the test in the natural climate varied between 10 and 30°C, the relative humidity between 40 and 90%. The strength values, plotted in Fig. 11, were measured on aluminium bonds with shotblasted adherends, but it is certain that the general nature of the results is applicable to steel bonds. Under the higher temperature and the constant high humidity a faster degradation of shear strength is observed. A further increase in the degradation rate may be obtained if the specimens are stored in water at the same temperature as the artificial climate, as may be seen in Fig. 11(b). The residual strength measured on specimens aged under these conditions after four months exposure is about 30% of the initial strength and it seems very probable that in this state water diffusing from the environment is present in all areas of the lap joint. Thus, further degradation will occur. This can also be concluded from the results given in Fig. 12, showing butt-joint strength values, measured on mild-steel epoxy joints, stored in water of different temperatures.[22]

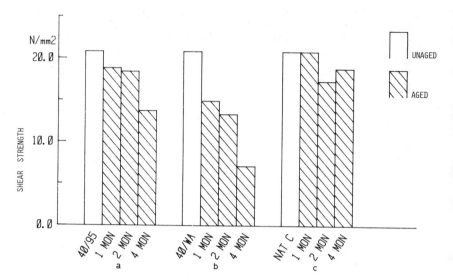

Fig. 11. Shear strength of aluminium joints. Adhesive, two-component epoxide resin. (a) Artificial climate, 40°C, 95% R.H.; (b) water, 40°C; (c) natural climate, North Germany.

The strength curve for the ageing condition at 40°C shows a high rate of degradation during exposure times between 1000 and 2500 h (2550 h = 3·4 months). Figure 12 also demonstrates the dependence of the degradation rate on the temperature of the environment. At higher temperatures the rate of degradation is greater but even after storing in water at 90°C residual strength values of 5 MPa were observed. The characteristics of all these curves may allow a prediction of lifetime, for example from a linear extrapolation of residual strength values, obtained after 700 and 1500 h of storage of a joint in environments of humidity higher than 70–90% R.H. or in water. However, the exposure times are too short for an exact derivation of a function for lifetime prediction.

In environments with relative humidity lower than *ca* 60% practically no degradation occurs; this is shown in Fig. 12, and is also the experience of the author, at least for times up to ten years.

The above comments refer to joints which were unstressed during ageing and it is even more difficult to give a reliable lifetime prediction of strength values for adhesively-bonded steel joints which in practice have to serve in humid environments under a sustained load. For

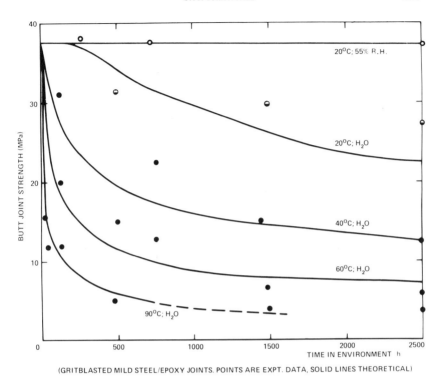

(GRITBLASTED MILD STEEL/EPOXY JOINTS. POINTS ARE EXPT. DATA, SOLID LINES THEORETICAL)

Fig. 12. *Butt-joint strength of mild-steel joints. Adhesive, epoxide resin.*[22]

example, the effect of hostile environmental conditions on steel joints exposed for times longer than one year are not generally available from the modern literature, which gives extensive information only concerning the long-term behaviour of aluminium bonds.[20] Nearly 18 years ago, however, Eichhorn *et al.* carried out detailed investigations concerning the long-term behaviour of bonded joints, from which some results are shown in Figs 13 and 14.[23] Plotted in these diagrams are the long-time shear strength curves measured on single-overlap stainless-steel joints under static loads in tap water at different temperatures. The adhesives were a one-component hot-curing epoxide resin (Fig. 13), and a cold-curing two-component methacrylate resin, cured by a peroxide hardener of the same type as the acrylic adhesive in Figs 9 and 10 (Fig. 14). If in these diagrams, for example, the strength

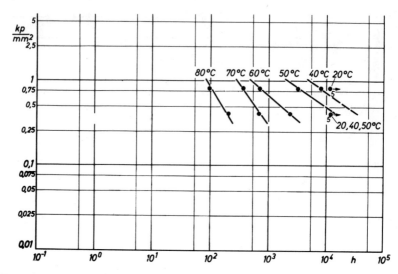

Fig. 13. Long-time shear strength of stainless-steel joints under static load in water at different temperatures. The stainless steel was sandblasted and nitric acid-etched. Adhesive, one-component epoxide–amide resin.[23]

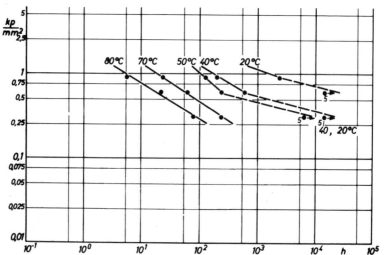

Fig. 14. Long-time shear strength of stainless-steel joints under static load in water at different temperatures. The stainless steel was sandblasted and nitric acid-etched. Adhesive, two-component methacrylate resin.[23]

versus lifetime curves measured in water of 40°C are linearly
lated until they cross the shear stress level of $0 \cdot 1 \, kp/mm^2$ (i.e. ,
ent to *ca.* $1 \, N/mm^2 \equiv 1 \, MPa$), then the intercept of both lines is
cases, located beyond a lifetime of $2 \times 10^5 \, h$, or about 23 years   ↩
lifetime is sufficient for most technical products. This simple calculation
leads to the same estimate of load capability of metal joints as that
evaluated from creep tests, discussed in section 3.2.2 of this chapter.

From these statements, scientifically inconclusive but of sufficient
reliability for the engineer, the following conclusion is possible: if the
design of an adhesively bonded steel joint, which has to serve under
humid environmental conditions, involves a level of sustained load
during its lifetime in the region of $1 \, N/mm^2$ shear, or about 1 to 5% of
the initial strength, then durability problems will be unlikely. This
does, however, assume that an *adequate surface treatment* is chosen
from short-time testing, for which results are given in Figs 9 and 10.
For the steel the surface treatment should usually be roughening by
grinding or shotblasting. The value $1 \, N/mm^2$ is a consequence of the
adhesive properties and assumes cold-curing adhesives are used, which
are very common in the steel industry. The value also takes into
account the moisture resistance of the adhesion between the polymer
and the metal surface. Finally, considering the properties of the
adhesive, then an increase in the long-term loading seems possible by
using hot-curing adhesives. Considering the adhesion aspects, then
other improvement techniques are possible, which will be discussed in
section 3.4.

3.3. Failure Mechanisms

As is discussed in the other chapters of this book, as well as in the
general literature,[24] in all polymer/metal bonds the stability of the
interfacial boundary zone is the critical feature for determining the
durability in aqueous environments. This is illustrated by the locus of
failure of initially well-prepared joints being cohesive in the polymeric
adhesive, but after long times in humid environments then so-called
interfacial failures occur. Indeed, macroscopic observation leads to the
conclusion that the adhesion forces between the polymer and the
metal, or the oxide layers on the metal surface, seem to have been
destroyed by the ingress of water. However, a more precise investigation of the fracture surfaces of such failures often shows that the failure
occurred either in the oxide layer on the metal or in an adhesive
boundary layer very near the metal surface. In the first case, small

amounts of metal or oxide are found on the adhesive side and in the second case small amounts of adhesive remain on the metal surface. In the case of adhesively-bonded aluminium joints, it has recently become increasingly certain that the so-called interfacial failures after ageing occur, in the case of etched aluminium, as mixed fractures in oxide and polymer. This can be seen from ultra-thin cross-sections using an electron microscope. If the same technique is used in delaminated bonds in which the aluminium surfaces were anodised, then the main failure mechanism may be a cohesive failure of the adhesive or the primer in a region near the oxide surface, which appears to be a weak boundary layer influenced by the chemical reactivity and the morphology of the aluminium oxides.[25] In the case of adhesive/steel bonds, the existing results from investigations of fracture surfaces are not so comprehensive, so an absolutely positive conclusion as to whether interfacial failure really occurs is not currently possible.

However, investigations have shown that in many cases of apparently interfacial failures between organic paint or adhesives and treated steel surfaces, small polymer residues did remain on the metal side and no metal or oxide, or only very small amounts, were found on the polymer side of the delamination.[26] If polymer remains on the metal surface the conclusion is therefore that within the polymer near the interface a weak boundary layer exists. This may be caused by water diffusing into this region of the polymer earlier than into the bulk polymer but the nature of this weak boundary layer within the polymer must also be influenced by the state of the metal surface on which the polymer is applied and, in the case of structural joints, cured. Indeed, results[26] obtained by means of ESCA showed a difference in the chemical properties between the remaining polymer residues on the metal surface and the bulk polymer. Thus, it is evident that the state of the polymer near the boundary layer in these cases must be influenced by the surface or, stated another way, the stability of the boundary layer can be influenced by changing the surface state by different treatments. This may arise from (i) selective adsorption by chemical means or (ii) the size of the micropores in the oxide layer. These may result in slightly changed curing mechanisms in the polymer boundary zone. However, such a mechanism still has to be conclusively established.

The influence on the polymer in the boundary layer of the state of the metal surface on which the polymer is cured, can be demonstrated

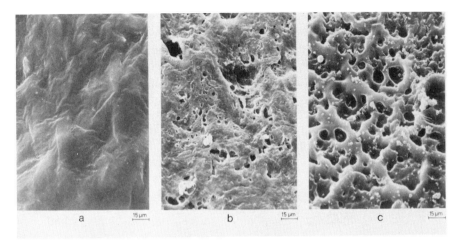

Fig. 15. Scanning electron micrographs of adhesive surface cured on mild steel subjected to different pretreatments; the steel has been removed by etching. (a) Sandblasted; (b) ground; (c) degreased. Adhesive, AF 42 (one-component hot-cured epoxide resin).

by experiments. If, for example, a one-component epoxy adhesive is cured on mild steel subjected to different surface treatments (e.g. shotblasting, grinding or only degreasing) and, after curing, the metal parts are etched away in a dilute aqueous solution of nitric acid, then only in the case of shotblasted steel is a true replica of the surface morphology obtained. This is shown in Fig. 15(a). Alternatively, if the one-component adhesive is cured on ground or degreased-only steel, the etching process produces a porous structure on the adhesive surface. The importance of such investigations is that they demonstrate that the stability of an epoxide adhesive to an etching solution depends upon the state of the surface against which the adhesive was cured.[17] Although it is still too early to give more details about these phenomena in a boundary layer, one conclusion may be of importance for future work on optimising the interface for good durability: if a weak boundary layer within the adhesive is a reason for the so-called interfacial failures it may be possible to optimise an adhesive/ metal interface, and particularly an adhesively-bonded steel joint, not only by changing the surface treatment but also by changing the structure of the adhesive in the boundary zone.

3.4. Techniques for Increasing the Long-Term Stability

Nearly all attempts made in the past to increase the stability of the interfacial boundary zone in metal/polymer bonds have been more empiric than scientific. This is because knowledge about reactions in the boundary zone, especially those at the interface between the adhesive and the metal surface, is relatively poor. This is also the case for the fracture mechanisms after ageing processes. However, in recent years more systematic tests have been carried out. Mostly they have started from the realisation that, whilst the bonds arising in the boundary zone between a prepolymer (or a macromolecular polymer) and a metal surface are of high strength, they are not necessarily stable against water. This is easy to demonstrate by adsorption tests on metal surfaces commonly prepared and used in industry or by chromatographic investigations on metal oxides.[27] The results of these experiments have led to the conclusion that invading water in the glueline can also destroy the chemical bonds in the boundary zone, for example, by hydrolytic reactions. If such a mechanism is supposed it should be possible to increase the stability of the boundary zone by replacing the originally water-unstable chemical bonds by stable systems. This can be produced, for example, by using substances which react with the metal oxides to more water-stable chemical compounds and, after this, react via other chemical groups with the adhesive to form covalent bonds. Classical substances which can react more readily with the metal oxides, or with the polymers, and build up water-stable bonds are reactive organic silanes. They have been used for many years as finishes on glass fibres and the matrix resins in fibre-reinforced plastics.[28]

If such a primer, based upon a reactive silane, is used for example in mild-steel bonds, a marked improvement in joint durability can be achieved, as shown in Fig. 16. Of some importance may be the fact that reactive silanes, like epoxy- or amino-silanes, increase the durability between the adhesive and metal surface not only if they are used as primers directly in the boundary layer, but also if they are mixed into the adhesive before the bonding process. Indeed, in some cases, this latter method is more efficient with respect to durability. One result of such tests is shown in Fig. 17, in which the shear strength between wood and steel is plotted. The ageing of these shear specimens involved exposure for 164 h in water at 40°C. The water in these tests contained 0·5% detergent, which increases the deterioration of the adhesion drastically, especially in wood/metal bonds. The adhesive

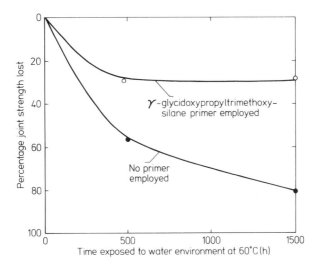

Fig. 16. Loss of strength of gritblasted mild-steel joints. Adhesive, epoxide resin.[28]

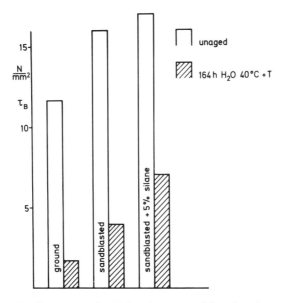

Fig. 17. Shear strength of joints between mild steel and wood.

used was an epoxide–aminoamide cold-curing system. The residual
strength was markedly improved by mixing 5% aminosilane with the
adhesive.

The positive effect of mixing a reactive silane into the adhesive is *not*
due to the permeability coefficient and diffusion constant of the epoxide resin being decreased by the additive. Measurements of water
diffusion in adhesives with and without reactive silanes showed that a
silane content of *ca* 2% did not change the diffusion properties of the
epoxide resin. On the other hand, this does not exclude the possibility
that the silane content within the adhesive causes positive effects in a
supposed weak boundary layer which is produced by influence of the
state of the metal surface.

Besides silanes, there also other organic substances which can react
with the adhesive to form water-stable chemical bonds with metal
oxides in the form of chelate complexes. Used as primers, for example
on aluminium surfaces, they lead to relatively small improvements of
the water-stability of the bonds. However, if they are mixed into
adhesives, for example two-component epoxide systems, improvements are observed in long-term durability, as may be seen in Fig. 18.
The improvement in the residual strength arises after ageing times of
more than eight months in a humid artificial climate. This represents
the time after which deterioration processes in the boundary layer start
to dominate the behaviour of the joint, rather than any changes in the

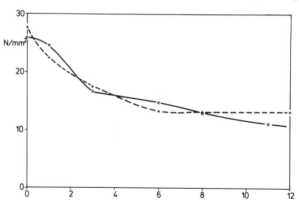

Fig. 18. *Shear strength of sandblasted mild-steel joints and dependence on
ageing time (in months) in a humid climate (40°C/95% R.H.). Adhesive,
Klima 40/95 (two-component epoxide resin). +−−−−+, St37 with alizarine;
O———O, St37 without alizarine. (St37 is a mild steel of tensile strength
37 kg/mm².)*

adhesive properties due to water diffusion. So, despite the fact that the chelate complex was mixed with the adhesive, the most probable mechanism by which the durability is improved is enhancement of the strength of the boundary layer. It is noteworthy that substances like alizarine, salicylic acid and others also improve the durability, if they are used in adhesive joints for aluminium and titanium.

The examples given above show the possibility of improving the durability of steel joints produced by organic adhesives in a more systematic manner than was previously adopted. The reaction mechanisms of the substances mentioned are not yet fully understood but the results demonstrate that further improvements in the durability of adhesively-bonded steel joints may be expected in the future.

4. SHORT-TIME TEST METHODS

All results concerning the durability of bonded steel joints reported so far involve tests needing expensive equipment such as climatic chambers and testing machines and well-trained personnel to carry out the experiments. In most factories or production lines in which steel joints are produced it is not usually possible to carry out such complex experiments simply to control the quality of the surface pretreatment process or the bonded joints. However, as pointed out in earlier chapters, a relatively simple short-time test method has been developed, namely the wedge test. The specimens in this test are cut from two metal sheets bonded together. In the USA, sheets of 3·2 mm thickness are common, but in the experiments carried out by the author on the durability of steel joints, sheets only 1·6 mm thick were used for this test method. After bonding, the metal plates were separated by driving a steel wedge of 4 mm thickness into one end of the adhesive layer. This wedge produces a crack in the unaged specimen for some millimetres down the adhesive layer, which is easy to observe from the edges. The wedge between the plates produces high stresses at the tip of the initial crack in the adhesive layer. The specimens are next exposed to a hostile environment, such as a humid climate, for several hours. The dependence of the water stability of the adhesion during this ageing process is illustrated by the crack propagating from the middle of the glueline into the boundary layer and doing so, in the case of unstable adhesion, at a relatively high velocity. In the case of

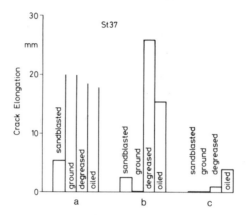

Fig. 19. Crack extension in mild-steel wedge specimens and dependence on surface treatment. Adhesives: (a) Two-component cold-cured epoxy; (b) two-component acrylic: (c) one-component hot-cured epoxy. Ageing process, 48 h. Artificial climate, 30°C/95% R.H. (St37 is a mild steel of tensile strength 37 kg/mm².)

stable adhesion it propagates only within the adhesive and only over a small distance of between 0 and *ca* 8 mm.

Some results obtained by this test method are plotted in Fig. 19. The adhesives used were a polyaminoamide–epoxy resin, a styrene–peroxide-cured acrylate, and a one-component epoxide resin. The mild-steel substrates were pretreated using different procedures. Plotted in the Figure are the crack lengths, measured on the sides of the specimens after an exposure time of 8 h in an artificial humid climate of 30°C and 95% R.H. It may be readily seen from the crack elongation that the surface treatment has a major influence on the durability behaviour. Compared with the results obtained from a classical ageing test on single-overlap joints (Fig. 9) it is clear that the wedge test leads to results of the same general character. If it is not possible to measure the crack length before and after storing the specimens in a humid environment, in most cases it is sufficient to initiate the crack without measuring the crack length, expose the specimens in the humid environment for some hours and then separate the joint by mechanical force. If the initial cohesive crack has not changed into an apparent interfacial mode during the exposure time, then the water stability of the adhesion is sufficient for most purposes.

The disadvantage of this test method is that the results are only of a qualitative nature, so it is not possible to distinguish quantitatively between different grades of interface stability. If such a measure is needed, another test procedure is required. The peel test is useful in this context. In this test the initial crack, generated in a dry environment, is stopped after peeling a length of some centimetres. Then, at the tip of the crack, a drop of water containing 0·5% detergent for better wetting is applied and, now in the wet state, the peel process is continued. The time to initiate an interfacial failure in such a peel specimen, in the case of unstable adhesion, is no longer than a few seconds. Thus, if the water stability of the adhesion forces is insufficient, the peel strength decreases rapidly and an interfacial failure occurs. Only in the case of very good and stable adhesion is there no change in the peel strength and no interfacial failure, or only partial interfacial failure, is observed. From such tests conducted in the author's laboratory it has been found that this wet-peel test appears to be more reliable than the wedge tests, and that the detection of unstable adhesion is better. Further initial results[29] from investigations by the aircraft industry have led to the conclusion that there is a direct connection between the behaviour of a joint in the wet-peel test and the ageing behaviour in a natural humid climate.

Both test methods lead to results after minutes or some hours, and so they are useful tools for quality control in many areas. They need not only be used for the testing of adhesive-bonded joints, but can also be used for the testing of the adhesion between coatings and paints on metal surfaces. In these circumstances the coating or paint is simply used as the 'adhesive' in the test specimen.

5. EXPERIENCES WITH BONDED STEEL JOINTS

To give an overview of the wide area of successful application of adhesive-bonding technology in industry would lead to a very long chapter. However, it is useful at the end of a description of experimental results concerned with promoting the failure of adhesively-bonded steel joints under hostile environmental conditions to make some comments about the real behaviour of bonded joints. Such joints often survive for more than 20 years despite the possible detrimental mechanisms which may occur over these long periods, and despite the harsh environmental conditions in which they frequently operate.

At the beginning of this chapter, it was stated that the metal bonding technique is about 40 years old. This is true for the case of structural adhesives which are applied as low molar-mass liquids and cured in the adhesive joint by a crosslinking process. More generally, bonded metal parts have been used by men for more than 200 years. Good examples are leather-coated watch cases, dating from the seventeenth and eighteenth centuries, in which the leather coat was bonded, especially those made by the famous English watch-making industry. Such watches are available today and the bonds between the leather and the metal case are mostly intact. Another example of an early application of adhesive bonding technology is the glass shield and reflector of headlamps of a small sports car, built in Czechoslovakia in 1933. The bonded part is shown in Fig. 20. The adhesives of this period consisted of rubber solutions or asphalt mixtures and, as the example in Fig. 20 shows, the bonded joints have served for about 40 years under severe conditions without debonding between the adhesive and the metal parts being detected.

Good experiences with bonded steel parts have also been recorded in the modern car-building industry where, especially in body construction, adhesives have been used successfully for more than ten years.

Fig. 20. Glass shield bonded on a headlamp reflector. Service time, 1933–1972.

Fig. 21. Bonded stiffener frame in the boot-lid of a car.

One example is shown in Fig. 21, in which a stiffener is bonded inside the boot-lid of a modern car. In this field the plastisol systems mentioned before have been used, since they need no special surface treatment and thus it is possible to apply the adhesive on the oily body sheets, assemble the parts, degrease, phosphate and paint the surfaces in subsequent steps. Finally the adhesive is gelled or cured together with the paint being stoved. More recently plastisol adhesives have been replaced in some areas by modern one-component epoxide adhesives, which also show good adhesion on surfaces without needing pretreatment. The car-building industry also uses bonding techniques for fastening windscreens into the body frames and for joining mirrors to the windscreens. Somewhat different uses are in the engine, where adhesives have served in the presence of water, in water-pumps, over several years. Screws are fastened by adhesives today in many important parts such as brakes and steering assemblies without any problems. In all, a modern car contains more than 5 kg of adhesives, and due to the good experiences with adhesive joints, the interest in replacing more and more welded or screwed constructions by bonded joints is still increasing.

Further examples of bonded systems are those which have served on the West German railways for many years. Modern rails are normally welded together to obtain better driving characteristics of the trains because, in winter, screwed joints produce gaps between the rails due to the contraction of the steel at low temperatures. This is shown in

Fig. 22. Screwed rail joint at −28°C. By courtesy of Henkel.

Fig. 22 for a screwed rail joint at −28°C. However, in modern railway systems with inductive security devices to send electric impulses from the track to the driver's cab, the rails from time to time need insulating joints. If they are screwed with lashes and isolating materials between the lashes and the rails, then they show the same contraction effects. This can, however, be avoided by a combination of bonding and screwing techniques in these joints. Cold-curing adhesives are used which are applied between the lashes and the rails, the surfaces of

Fig. 23. Screwed and bonded rail joint at −28°C. By courtesy of Henkel.

which are either ground or shotblasted. Such a screwed and bonded rail joint at a temperature of $-28°C$ gives no gap, as may be seen in Fig. 23. Comparison of Figs 22 and 23 makes it clear that the adhesive in addition to the screws transmits a large part of stress in this construction. The experience with bonded and screwed rail joints is therefore very good and demonstrates that, despite some environmental attack observed from laboratory tests, the application of adhesive bonding in joining steel parts can be very useful and successful.

REFERENCES

1. Preiswerk, E. and Zeerleder, A. V., *Schweiz. Archiv für angewandte Technik*, **12** (1946), 113.
2. de Bruyne, N. A. and Houwink, R., *Adhesion and Adhesives*, Elsevier, New York (1951).
3. Golf, W. E., *Aircraft Production*, **8** (1946), 211.
4. Trittler, G., *VDI-Nachrichten*, **17** (1963), 325.
5. v. d. Laden, E., *VDI-Nachrichten*, **14** (1960), 1.
6. Hinterwaldner, R., *Adhesives Age*, **21**(8) (1978), 34.
7. Wangman, C., *Adhesives Age*, **20**(9) (1977), 23.
8. Mittal, K. L., in *Surface Contamination, Genesis, Detection and Control*, Vol. 1, ed. K. L. Mittal, Plenum Press, New York (1979), p. 3.
9. Martin, J. T., in *Adhesion and Adhesives*, ed. R. Houwink and G. Salomon, Elsevier, Amsterdam, London, New York (1967).
10. Brockmann, W., *Adhäsion* (1969), 345, 448; (1970), 52, 250.
11. Gierenz, G., *Referate* No. 5, Henkel, Düsseldorf (1970).
12. de Bruyne, N. A., *Aircraft Eng.*, **16** (1944), 175.
13. Goland, M. and Reissner, E., *J. Appl. Mech.*, **11** (1944), A 17.
14. Matting, A. (ed.), *Metallkleben*, Springer, Berlin, Heidelberg, New York (1969).
15. Hart-Smith, L. J., in *Developments in Adhesives 2*, ed. A. J. Kinloch, Applied Science Publishers, London (1981).
16. Althof, W., The diffusion of water vapour in humid air into the bondlines of adhesive bonded joints, *Proc. 11th Nat. SAMPE-Conf. Azusa, California* (October 1979).
17. Brockmann, W. and Kollek, H., Ermittlung der Langzeitbeanspruchbarkeit von Metallklebungen mit im Maschinenbau gebräuchlichen und zukünftig verwendbaren Bindemitteln, *FKM-Forschungshefte No. 81*, Maschinenbau-Verlag, Frankfurt (1980).
18. Althof, W. and Brockmann, W., *Adhesives Age*, **20**(9) (1977), 28.
19. Steffens, H.-D. and Brockmann, W., *Aging Resistance of Light Metal Bonded Joints Using New Surface Treatment Methods*, Royal Aircr. Establ. Library Transl. No. 1662 (1972).
20. Minford, J. D., *Int. J. Adhesion Adhesives*, **2** (1982), 25.

21. Brockmann, W., *Adhäsion,* **17** (1973), 72.
22. Gledhill, R. A., Kinloch, A. J. and Shaw, S. J., *J. Adhesion,* **11** (1980), 3.
23. Eichhorn, E. *et al.,* Untersuchungen über das Alterungsverhalten, die Temperaturabhängigkeit und Zeitstandfestigkeit von metallklebverbindungen mit und ohne Füllstoffzusätze zum Klebstoff, *Forsch. Ber. Land Nordrh. Westf. No. 1734,* Westdeutscher Verlag, Köln u. Opladen (1966).
24. Kinloch, A. J., *J. Mater. Sci.,* **17** (1982), 617.
25. Hennemann, O-D. and Brockmann, W., Weak boundary zones in metal bonds, *Proc. 14th SAMPE Conf., Azusa, California* (October 1982), p. 302.
26. Hercules, D. M., *J. Electron. Spectrosc. Rel. Phen.,* **5** (1974), 811.
27. Brockmann, W., *Adhesives Age,* **20**(7) (1977), 33.
28. Gettings, M. and Kinloch, A. J., *J. Mater. Sci.,* **12** (1977), 2049.
29. Knock, K. K. and Locke, M. C., Correlation of surface characterization of phosphoric acid anodized oxide with physical properties of bonded specimens, *Proc. 13th Nat. SAMPE Conf., Azusa, California* (October 1981), p. 445.

8

Aerospace Applications

P. ALBERICCI

British Aerospace, Woodford, Cheshire, UK

1. WOOD AND FABRIC AIRCRAFT

Self-sustaining powered flight was first accomplished by the Wright brothers at Kitty Hawk, North Carolina. They flew for 12 seconds or so and covered around 150 feet. Technical progress was slow for the next few years but speeded up noticeably around 1910, by which time the Channel had been flown and several military authorities were beginning to take a serious interest in aviation. The outcome of this military interest was to be seen a few years later when hundreds of fighters, observation machines and light bombers faced each other along the length of the Western Front. But not only along the Western Front: by 1917 Zeppelin raids over London and the East-coast towns were commonplace and lighter-than-air 'airships' 650 ft long and operating at heights up to 20 000ft were carrying the war very many miles from their home bases. What is not generally known, however, is that nearly half of the 'Zeppelin' fleet were not Zeppelins at all—they were craft built by the competing firm of Schutte–Lanz and they differed from their better-known counterparts in two important respects—they featured a lower drag aerodynamic form which made them faster, and their structures were built of wood in total contrast to Count Zeppelin's designs which pioneered the use of the new German high-strength aluminium alloys, later to be known throughout the world as Duralumin.

The Schutte–Lanz airships used spruce and ply box beams and wooden lattice girders in their construction and glue was utilised extensively in the structural assembly. The glues available at the time were all derived from natural proteins, well-known examples of which are the animal bone and skin glue, fish glue, and gelatine. These are worked as thick, high-solids solutions and set solely by diffusion and evaporation of water through the wood. No irreversible chemical change (as in urea–formaldehyde adhesives) or physical change (as in emulsion types) takes place and joints made from these glues rapidly lose strength and fail in damp conditions. The Schutte–Lanz designers were however aware of a newly developed milk casein material with very useful properties as a wood glue. Casein as prepared is quite insoluble in water but can be dispersed in dilute alkalis such as calcium hydroxide and borate solutions. When set a moderate degree of water resistance is obtained. This degree of water resistance was not enough, however, for the Schutte–Lanz airships operating in winter in the moisture-laden air over the North Sea. Glue weakening was inevitable and structural failures became commonplace as it was quite impossible to protect the vast area of glued wooden structure from the damp air entering through the thin outer fabric envelope. Even a small degree of plasticisation by water uptake can result in a large loss in joint strength, especially when high internal stresses are fed into the joint system through moisture-induced expansions in the wood. A general lack of confidence in the glued wooden airship had set in by the second half of the war and further development of the Schutte–Lanz designs was abandoned, the two firms amalgamating before the end of hostilities. All the later German airships—the 750 ft-long super Zeppelins—really were Zeppelins continuing the successful use of the new high-strength aluminium alloys being developed in Germany.

Airships apart, wood continued to be widely used in the production of aeroplanes and although it is possible to utilise wood in aircraft structures without resource to adhesives at all (through the use of metal fasteners and fittings) the most efficient use of wood—which features highly directional mechanical properties—requires the adoption of a load-spreading adhesive. Streamlined, monocoque shell construction (as opposed to square-section stick and fabric) depended absolutely on the use of adhesives for the large-area bonding involved. Whereas no allied World War I aircraft adopted streamlined monocoque wooden construction the adhesive-bonded wooden-shell approach was popular amongst German designers. Their Albatross, Pfalz, and Roland machines all featured stressed-skin wooden fuselages whose birch veneers were wrapped and planked in various configurations but invariably were bonded together, and to their frames, with casein glue. As these aeroplane structures were quite compact they were effectively protected from the weather and other damaging agencies by varnish on the inside and cellulose-doped fabric on the outside, in spite of the frightening fire risk associated with nitrocellulose. Thus protected, the combination of wood and casein proved both practicable and successful and continued to be used, particularly in civil aircraft, for many years. It should be remembered, however, that life expectancy of any aircraft was really only a few years (through technical obsolescence) and the aircraft were normally kept in a hangar when not flying, which is not the case today. Interestingly, laminated wooden propellers were generally bonded with traditional carpenter's animal glue because at the normal ambient water content of wood it is stronger and tougher than casein. External protection was of course paramount but this could be readily achieved through multiple coats of varnish or through the application of a solvent-softened celluloid (nitrocellulose again) sheathing. Laminated and densified wooden propeller blades fitted to complex variable-pitch hub mechanisms were widely used during World War II and competed keenly against light alloy and steel counterparts. High rates of production were the order of the day and by 1940 two very different types of 'modern' synthetic resin adhesives were being used. One major manufacturer used thermosetting phenolic resin as the bonding agent whilst the main competitor used the thermoplastic resin poly(vinyl formal) for the same role. Although totally different in their chemical and mechanical nature both adhesives gave complete satisfaction throughout the whole period the densified wooden bladed propeller was in use. With the emergence of the final generation of piston engines (2000 hp and more) wooden blades, however densified, could not handle the concentration of load at the hub and and their development ceased. Throughout the several years of intensive war-time use no problem whatsoever was experienced with either type of these hot-press bonded laminated propellers, but again it has to be pointed out that the wood itself required very elegant forms of surface protection, particularly at the leading edge, so that in practice the actual bondlines were not really exposed to the weather.

Returning to the airframe itself, two of the most successful passenger-carrying machines operating during the early years of civil flying were the Fokker tri-motor built both in Europe and in America and the smaller Lockheed Vega designed and built in Hollywood, California. The Fokker tri-motor featured the welded steel tube fuselage (fabric-covered) and all-wood cantilever wing combination which served so well in the World War I Fokker fighters (they replaced the more elegant-looking but slow-to-produce wooden monocoque fuselage types mentioned earlier). The smaller Lockheed Vega had a wooden cantilever wing of generally similar construction but differed completely in its fuselage design, which was unique in being a highly efficient (both structurally and aerodynamically) wood-veneer monocoque shell which was pressure-bag moulded in two halves using a concrete mould. This was the first attempt (*ca* 1927) at pressure-bag moulding of large structures—a manufacturing technique which was to re-emerge over 20 years later with the appearance of room-temperature curing, liquid form, epoxy and polyester laminating resins and the various high-strength fibres. Fokker and Lockheed were however limited to casein for bonding their wooden structures and whilst initially very satisfactory bonding was achieved, long-term pan-climatic operations, by then becoming commonplace, began to under-line the water sensitivity of casein glue (casein 'cement' as it was frequently called). A dramatic end to wooden 'airliners' in America came in 1931 with the highly publicised crash of an American-built Fokker tri-motor. Structural failure of the wooden wing was the accepted cause and the event led to the complete withdrawal of wooden-structured aircraft from the fare-paying passenger traffic scene on the American continent. It was never made clear whether failure of a wooden member or failure of the adhesive was the actual cause of the crash. Within a few years aluminium-skinned all-metal aeroplanes dominated the commercial scene completely, although experimental work on the pressure bag (or autoclave-moulded) wood-veneer technique continued for many years for small private aircraft using heat-curing phenolic resins as the bonding agent.

Although abandoned elsewhere, the De Havilland company in Great Britain continued to pursue the use of glued wood for large commercial aircraft and developed a new and highly efficient form of 'sandwich' construction (birch veneer skins and balsa core) for the superbly streamlined DH 91 Albatross 4-engined airliner put into service by Imperial Airways at the start of 1939. These aircraft were stored in the open when civil operations ceased at the start of World War II. Put into war-time communications service about one year later a number of minor structural 'happenings' disclosed that the casein glue had deteriorated and was unbonding in a number of localised areas, and the type was subsequently withdrawn from further service.

The advanced highly efficient birch ply skin/balsa wood core form of construction used for the DH 91 airliner was carried forward by the DH company and was adopted in its entirety for its DH 98 design—the famous Mosquito, one of the most outstanding aircraft of the Second World War. Entering RAF service in 1941 early examples of Mosquito were built using casein as the assembly adhesive but the bonding agent was changed to the acid-hardened urea–formaldehyde (UF) type as soon as the new, wholly synthetic, cold-setting adhesive became available in the quantity required by the scale of production. Nearly 8000 of these all-wood warplanes were built (in the UK, Canada and Australia) and they saw service in every theatre of war. Only in the Far Eastern theatre was trouble experienced with the glued wooden structure. In this respect whilst well-cured UF adhesive is more than adequately resistant, extreme limits of humidity, let alone contact with water, cause the wood itself to become highly unstable dimensionally. In particular it undergoes massive movements in the cross-grain direction. It is these dimensional changes in the wood with changes in humidity or actual contact with water which pose the problem. The situation is all the more difficult in the aircraft case as the structural sections are relatively thin so that the moisture content of the wood rapidly equilibrates with its surroundings and changes dimensions accordingly. The hard brittle nature of the densely crosslinked urea and phenolic type resins does not make them sympathetic to the stresses set up by the movement in the cross-grain direction of the wood. Whilst short-term protection can be achieved by the use of surface coatings, success in the long term is too difficult in practice because of imperfections in the coating or else because of the weight penalty of thick coatings.

A little-known sequel to the Mosquito success story can be found in the history of the Focke–Wulf TA 154. The existence and subsequent technical excellence of the British wooden warplane made a great impact on the German authorities and a very similar counterpart, also made in wood, was rushed through development and into production. Unofficially dubbed the 'Moskito', two prototypes crashed during trials in 1944, one shortly after the other, breaking up in the air. The cause of the crashes was found to be weakening in the wood adjacent to the bonds brought about by the use of a resin-hardening agent which was too strongly acidic. Only a handful of these aircraft entered service and the type was regarded as a failure. It is possible that a cold-setting phenolic type resin was used in this case as tendering of the wood leading to rapid joint failure was also encountered in the few cases where cold-setting phenolic resins were used in this country as an assembly adhesive for wood. It is somewhat ironic that the Germans, who were the recognised leaders in theoretical resin chemistry, should have so failed in practical application in the field. It should also be said that the Focke–Wulf Company who designed the aircraft had no previous experience in the manufacture of wooden airframes, unlike the De Havilland Company whose experience with wood was unrivalled.

A large number of wooden-built trainer and communication aircraft came on to the civil market in the immediate post-war period. Aircraft built by Miles, Percival, Airspeed, De Havilland, and the early version of the Avro Anson which had a wooden wing, were all very suitable candidates for peace-time civil flying. However, few of these types remained in service for any length of time because of the discovery of bondline deterioration within the hollow box type main spar structure. The closed-in box is attractive structurally because it combines torsional stiffness with bending stiffness but it suffered from the endemic corrosion problem of all box or tube structures. They are rarely completely closed, so that damp water-containing air readily enters depositing condensed water at low temperatures which may take weeks or months to escape again. Those types featuring hollow box type spars were difficult to inspect economically as required by the airworthiness regulations and so they rapidly faded from the aviation scene.

Wooden sportplanes and gliders continue to be made in small numbers up to the present time using mainly UF adhesives but some casein was used in France as late as 1960. Longevity of the structure still depends, however, on maintaining a benign ambient humidity and keeping any free water (from rain or condensation) completely outside the aeroplane. Strangely, the type of wood adhesive which is universally regarded as being the most water-resistant and durable—the resorcinol–formaldehyde type—has received very little attention from the aircraft industry. Mainly this is because the setting time is too slow (it is used in the essentially neutral state without acid acceleration) but also because the real need is not so much for added durability for the adhesive but for greatly increased flexibility. The emergence of crack-stopping toughening modifiers for the too-brittle adhesives came too late for the wooden aeroplane.

2. STRUCTURAL ADHESIVE BONDING IN METAL AIRCRAFT

2.1. Historical Development

High-strength metal glues, generally called structural adhesives, have been used in the aircraft industry for well over 30 years and it is worth recalling that this daring concept was very much a British development in the initial stages. In order to understand better the problems of reliability and durability associated with the rise of metal adhesives the following list of 'milestones' is offered, to illustrate the size and scope of the technology involved. For those readers not associated with the industry the main areas of application are depicted in Figs 1–4. Note that the *raison d'être* behind the stringer/skin and sandwich construction (wood or metal) is the minimal-weight stiffening of the thin outer skin or shell so that it does not buckle and collapse under compression or shear loads—for example, in flight the top surface of a cantilever wing is severely loaded in compression.

1942　'Redux' vinyl formal/phenolic metal adhesive pioneered by De Bruyne and Reyner. Parallel developments in USA.

1944　First production application. Redux bonding of alloy spar booms to wood in DH Hornet (successor to Mosquito) twin-engined long-range fighter.

1945　Metal-to-metal bonding (stringer/skin) extensively used in DH Dove feeder-line passenger aircraft. Adhesive used, Redux, bonded via wide-bed hydraulic press.

1950　Metal skin–balsa wood core sandwich appears in American Chance Vought Cutlass naval fighter. Adhesive: elastomer–modified phenolic.

1952　Jet-powered airliner DH Comet in service (Fig. 5). Metal bonding as in Dove but extended even to double curvature areas of fuselage. Metal bonding considered very daring feature for jet aircraft.

1955　Metal bonding used by Continental manufacturers. Fokker F27 Friendship uses Redux in generally similar manner to that employed in the Dove (1945) and naval fighter (1950) described above. Very popular aircraft also built in America under licence.

1956　Aluminium skin, aluminium foil honeycomb sandwich construction developed by Martin and Hexel Corporation in America.

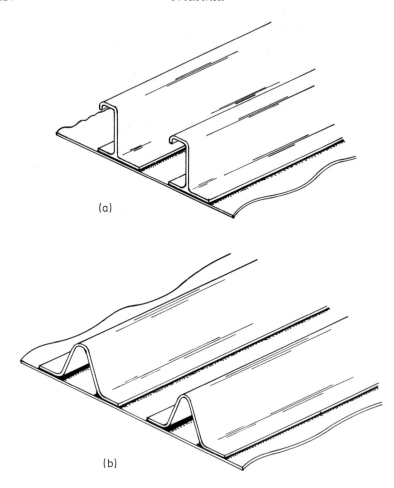

Fig. 1. *Stringer/skin bonded joints. (a) Extruded J-section stringers; (b) rolled strip closed-channel stiffener.*

First major application on wing structure of jet-powered flying-boat, Martin Seamaster. Small all-honeycomb fighter built by Avro in UK. Epoxy adhesive for bonding skins to core.

1958   Integral fuel tank wing structure ('wet wing') by Convair uses film-type adhesive plus riveting for assembly and sealing. Adopted for F102 and F106 fighters and Convair 880 and 990 airliners. Adhesive–nitrile rubber modified phenolic film.

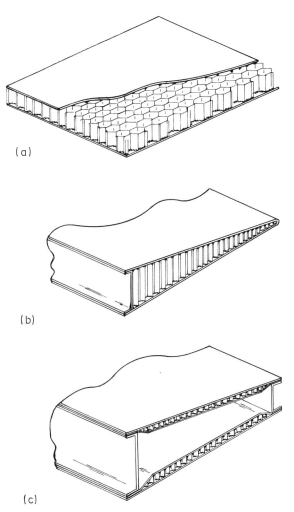

Fig. 2. (a) Honeycomb panel (flooring, etc.); (b) honeycomb structure (control surfaces, trailing edges, etc.); (c) honeycomb structure (aerofoils, etc.).

1960 Aluminium honeycomb sandwich used extensively in Convair B58 Hustler supersonic bomber (Fig. 6): High temperature epoxy–phenolic required to cope with service temperature of supersonic flight. Retired from operations in 1970.

1963 Cold-setting epoxy adhesive used by Boeing in 727 and other

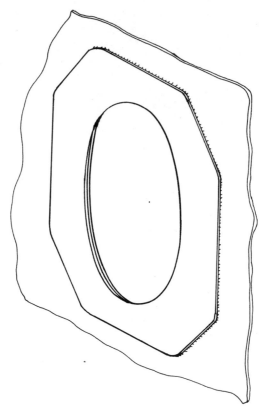

Fig. 3. Bonded doubler plate (window cut-outs, access holes, etc.).

airliners in pressure cabin joint interfay. Refrigerated adhesive placed in joint prior to riveting. Aim to increase fatigue life.

1965   Brequet 'Atlantic' multinational maritime reconnaissance turbo-prop machine (Fig. 7). Extensive use of aluminium, honeycomb sandwich construction for wing and fuselage. Modified-epoxy adhesive.

1969   Fokker introduces magnetic clamping to longitudinal fuselage joints to consolidate cold-set epoxy adhesive prior to riveting in process similar to that employed by Boeing (1963), but more sophisticated. Aim again improved fatigue life.

1971   Use of clad sheet ('Alclad') identified in various American reports as potential source of premature joint failure. Recom-

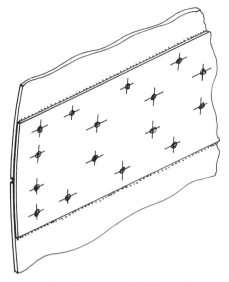

Fig. 4. Riveted and bonded lap joint.

mendation that clad sheet (widely used in aircraft) should not
be used for adhesive bonding.

1972   Lockheed Tri-Star introduces use of titanium crack-stopping
reinforcing straps in pressure cabin. Modified-epoxy film used
for bonding titanium to aluminium.

1974   'Weldbonding'—Russian Anatov An-24 reported to use a combination of electric spot-welding and adhesive bonding in its
construction. Details not known.

Fig. 5. De Havilland Comet I, 1952, incorporating daring and extensive use of structural bonding.

Fig. 6. Convair B58 Hustler supersonic bomber: epoxy-phenolic adhesive, with 4500 ft² of aluminium honeycomb sandwich, in the construction.

1975 US Government sponsored PABST program (Primary Adhesively Bonded Structures Technology) launched, aimed at extending use of adhesive bonding and also to optimise manufacturing procedures.

1976 Preferred surface pre-bond treatment for aluminium (in USA) now emerging as phosphoric-acid anodising on non-clad alloy sheet. Reverses previous trend towards chrome–sulphuric pickle, long held in highest esteem. European manufacturers undecided.

2.2. Service Experience—Metal-to-metal Lap Joints

The adoption of metal bonding in the manufacture of aircraft quite naturally involved a great deal of testing and evaluation, so much so in fact that it has evolved a technology in its own right. Most of this testing related to material selection and to the optimisation of the

Fig. 7. Brequet 'Atlantic'—a seafaring exponent of the aluminium honeycomb sandwich structure, with modified-epoxy adhesive.

many process variables. Treatment of the metal surface prior to bonding was quickly recognised as being of critical importance and this area was singled out for much of the experimental work performed. A vast amount of mechanical testing—lap shear and peel—was undertaken and the effects of temperature were thoroughly explored. Nor were environmental agencies ignored: the effects of the various fluids used in and around aircraft were investigated. The influence of humidity and corrosive conditions were also investigated—but not to the extent that might have been expected. This was not really surprising, the life expectancy of any aircraft in the 1950s could be measured in single figures and metal corrosion and related problems had not yet emerged as problem areas. Largely this was because the utilisation rate of piston-engined aircraft was very low and also because low operating heights and lack of cabin pressurisation meant that the problem of heavy internal condensation was still to come. In summation, the general feeling of the time was that adhesive joints capable of sustaining 2500 psi in shear and 30 lb in the standard peel test were joints for life. The full thrust of the development effort went towards increasing the initial joint strength, the strongest joint being unquestionably the best joint in all respects.

One report dated 1952 did, however, touch on the problem that was to come but the implications went unnoticed by the industry at large. Leigh and Savage in a Royal Aircraft Establishment (RAE) report[1] found that Redux (vinyl formal–phenolic) joints on both pickled and anodised aluminium were only slightly affected by exposure to hot/wet cycling conditions whilst epoxy bonded joints on pickled aluminium were severely weakened. Epoxy joints on an anodised surface performed in a similar manner to the Redux joints. The distinction between the two adhesive types in their behaviour to the surface treatment was not underlined and in any case Redux, the only metal adhesive in widespread use, was not selective so there were no grounds for concern. In 1954, however, there were the most serious grounds for concern over the whole future of metal bonding in aircraft. The Comet fleet was withdrawn from service after two aircraft were lost under most mysterious circumstances. The daring design feature of structural bonding was high on the list of possible causes of structural failure although it should be said that the basic viability of metal bonding rather than durability was the main cause for alarm. The subsequent, classical, report of the Court of Enquiry firmly established metal fatigue as the cause of pressure cabin failure. Had the findings been less forthright, or precise, metal bonding with synthetic resins would have remained for evermore an aviation 'skeleton in the cupboard'.

Following the complete exoneration of adhesive bonding in the enquiry (redesigned aircraft appeared as Comet 4 with thicker fuselage skins) there was no loss of confidence in metal bonding and the technique continued to gather acceptance. In the UK and Europe this acceptance has centred mainly around (at least until very recently) the vinyl formal–phenolic adhesive type whilst in America a much wider range of materials has been brought into use. Nitrile rubber–phenolic, epoxy–phenolic, epoxy–polyamide and various other forms of modified epoxy appeared on the market, generally in film form. Operational experience with the very many aircraft (of several types from different manufacturers) using Redux and the nitrile-phenolics has been consistently good over time-scales of 10–15 and even 20 years. The pretreatment and the post-bonding protection are clearly variables which play their part but even in the least optimised situation the modified phenolic has rarely, if ever, resulted in overt de-bonding in a situation which did not at the same time produce general metal corrosion in the affected area. Failures can occur and have occurred, trapped water contacting an unprotected or poorly protected joint edge over a period of several years results in corrosion and de-bonding, but the incidence of such events has been low. Summing up, the modified-phenolic type adhesive used with a range of surface treatments and operating under many differing environments has been shown to maintain bond integrity well in line with the life expectancy of the aeroplane in which it is used.

Experience on the other side of the Atlantic with bonded structures has been somewhat different and altogether much less satisfactory. Their surface pretreatment (generally 'Forest Products Laboratory (FPL) etch' in America) was virtually the same but it was more generally linked to an epoxy type adhesive. One in particular (epoxy–polyamide) was highly favoured because of the facile low-temperature (120°C) cure coupled to an exceptionally high peel performance. Unfortunately this otherwise very attractive film-form adhesive produced joints of very poor durability when used in conjunction with the then standard surface treatment—the chrome–sulphuric (FPL) etch. Even under quite benign humidity, say 60–70% R.H., these joints were failing in a matter of a few years whilst under direct water attack failures could be identified in a matter of months, when no additional protection was possible.

One American marque in particular suffered a history of poor bond durability which in no way endeared it to the many operators flying the type. The problem centred around the use of a cold-setting epoxy used in the manner of a jointing compound or sealant placed within the interfay prior to riveting up. Within a few years of typical airline operation a high incidence of incipient corrosion and unbonding was becoming apparent from within the bondline, particularly in areas where condensation and waterlogging was taking place. Because, in this application, the adhesive was being used in a purely supplementary manner aiding the standard mechanical fasteners there was never any real question regarding flight safety but the costs of additional inspections and, in some cases, local rectification led to a general lack of confidence regarding the on-site use of cold-curing epoxy. In fact, this widespread condemnation was a somewhat unwarranted generalisation arising from a rather special case. Certainly no great attention had been given to the surface treatment by the production engineers involved and the low-mobility reactive adhesive used was held under refrigeration up to the moment of assembly. Hence the surface wetting behaviour of this system must be regarded as being very suspect (see Chapter 1). At a later date, and with due regard to the special requirements of cold-curing systems, other manufacturers have adopted a similar bond-*cum*-rivet (cold set) assembly technique with complete long-term success.

Returning to the general picture, there was a realisation during the 1960s that adhesive-bonded joints were not necessarily made for life and many aspects of adhesive technology began to be re-assessed. The emerging generation of large transport aircraft was now expected to remain in service for 15–20 years, perhaps 25 years. Corrosion generally was a problem and perhaps the breakdown of bonded joints was merely an extension of the corrosion problem area. Various reports from the RAE,[2,3] the Atomic Weapons Research Establishment (AWRE),[4] Sandia Laboratories,[5] and many other workers showed clearly however that a specific problem of adhesive bond breakdown existed when aluminium joints were exposed to high humidity or direct water attack. This work led the way in pointing out that things were happening at the adhesive/metal interface that were quite different from what was happening in the body of the adhesive. Also, whilst bond failures detected in the field were generally well associated with attendant corrosion (because of the elapse of time between initiation and detection), work in the laboratory showed time and time again that bonded joints could fail leaving behind perfectly clean metal surfaces with no signs of corrosion in the ordinary meaning of the term, as discussed in detail in Chapters 1 and 3.

Many observers tended to complicate the issue by insisting that a form of stress corrosion was at work and that some level of applied stress was an essential feature of the mechanism. However the role of external stress as the primary factor is difficult to reconcile with the many instances of small test pieces coming apart in laboratory exposure trials under zero load conditions. Also, failures in the field did not show any permanent set or any other signs of high loading in the metal.

Because of the enormous number of variables operating when making a typical bonded joint (even a standard acid-etch treatment can produce a very wide range of results depending on the bath operating conditions and the metal alloy specification) it is an invidious task to attempt a graphical setting-out of the wet degradation problem. To a certain extent it may even be misleading but nevertheless Figs 8 and 9 are given with the aim of providing at least some datum lines. Note that results presented are in the nature of worst cases—the joint area is small, the environment is total immersion, and none of the surface treatments are optimised. The poor durability of epoxy type film adhesives on the chrome–sulphuric pickle surface is well brought out in Fig. 8 and the marked improvement on standard chromic-acid anodising appears in Fig. 9. The diagrams also show the generally higher joint strengths achieved by the epoxy family over the original Redux, which explains why designers were drawn towards this class for their newer aeroplanes. Comparing any one adhesive across the two Figures brings out the point made early in this chapter; the surface treatment giving the highest initial joint strength (the acid etch) is not the one that gives the best durability. Unfortunately for some, up to around the year 1970 most of the development work concerned with metal bonding did in fact concentrate on the attainment of the highest *initial* joint strength. Subsequently, of course, the whole emphasis has changed, particularly in America, and the advantages of porous anodising have become widely recognised and put into practice. This development will be discussed in section 3.1.

Test techniques have also changed and the current trend is to examine the behaviour of 1-in wide bonded specimens in the cleavage mode. Two types of specimen configuration are in use (see Chapters 1 and 5). The double cantilever beam (DCB) test specimen is made from two strips of $\frac{1}{2}$ in $\times$ 1 in $\times$ 12 in plate bonded together and then forced

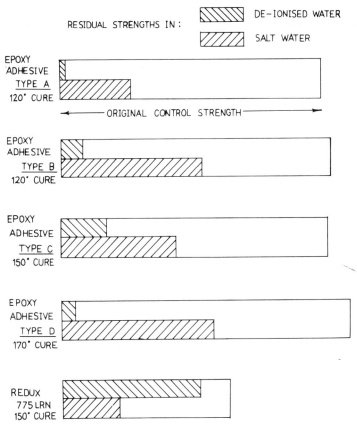

Fig. 8. *Loss in strength of aluminium/aluminium lap shear joints after 12 months in de-ionised water and in 3% salt water at room temperature. The aluminium (L72) was chrome–sulphuric acid-etched before bonding.*

apart so as to form a crack the length of which is monitored during the (relatively short) exposure period, frequently measured in hours. The cheaper wedge test, now established as ASTM D3762–79, uses $\frac{1}{8}$ in × 1 in × 6 in material forced apart by a 1 in × 1 in × $\frac{1}{8}$ in wedge. The standard exposure times are 1 h and 24 h under a 100% R.H. and 50°C environment. This new evaluation approach produces very rapid results and is very useful in direct 'back-to-back comparisons'. On the other hand the total system is very highly stressed in cleavage and

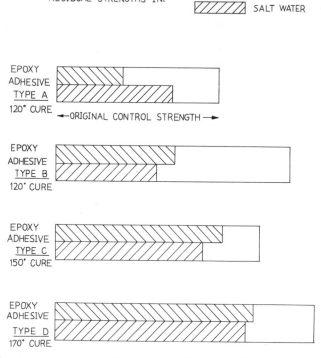

Fig. 9. *Loss in strength of aluminium/aluminium lap shear joints after 12 months in de-ionised water and in 3% salt water at room temperature. The aluminium (L72) was chromic acid-anodised before bonding.*

strongly favours those interface and adhesive combinations with highly developed crack-blunting and crack-stopping qualities. These features are certainly important to the designer but the very high short-time mechanical demands may well mask the long-term water-triggered reactions at the metal/adhesive interface which appear to be the real villain of the piece. Airline maintenance engineers would undoubtedly support this view—bond degraded sites are invariably associated with areas of high water concentration, not high stresses.

2.3. Service Experience—Aluminium Honeycomb Sandwich
Whilst the DH Dove and Comet and the Fokker Friendship introduced structural adhesive bonding to the airlines of the world it was the big

American four-jet airliners appearing on the scene in the early sixties which introduced aluminium honeycomb construction to civil operators. In general their experience with this later development was not as happy as with the first. Quoting a major airline Service Engineer,[6] 'Quite soon after the aircraft entered service, it became apparent that these various structures had certain drawbacks. One in particular was their susceptibility to corrosive attack on the bonded surfaces owing to the ingress of moisture via the edges of the skins. This was found to be common to the bonded skin panels of elevators, composite structure honeycomb flap panels and also all wedge-shaped trailing edge sections.' Ten years later and referring to a second-generation aeroplane, the head of maintenance and engineering for a cargo airline states:[7] 'Honeycomb sandwich is a material that is always attractive to the designers for weight reasons, especially in large areas, but which invariably, over the past two decades or more, has caused continuous maintenance problems'.

On the military aviation side the US Naval Weapons Bureau sponsored a major investigation into the water ingress problem of sandwich structure as early as 1959, but the problem remained intransigent and the US Air Force Materials Laboratory published the following statement in 1977: 'Although the structural performance of aluminium honeycomb sandwich is attractive, its durability performance in service (meaning principally resistance to moist environments) using early materials and process technology has not been as good as desired'.[8]

European experience has been very much the same as the American and this is only as expected as the techniques and materials of sandwich construction are virtually the same worldwide. The bipartisan differences referred to in section 2.2 have been less obvious in sandwich panel technology. A well-publicised victim of the sandwich syndrome was the multinational maritime reconnaissance aircraft which required much structural replacement after six or seven years of flying service. A major rebuild programme kept a large portion of the fleet non-operational to the embarrassment of various governments. Press statements at the time explained the situation by stressing the highly corrosive environment associated with over-the-water operations. The salt-laden atmosphere was singled out as the prime cause of the trouble and there would be some justification for this as some degree of corrosion as well as de-bonding would be present in the situation outlined. However, most of the world's major airports are situated on or near the coast and the salt-spray aerosol which is formed in storm conditions is rapidly distributed over the land mass by the self-same winds. It is really very doubtful whether maritime aircraft are subjected to any worse environmental conditions than the average civil transport. The carrier-borne naval aircraft is, however, a special case.

Technically the honeycomb sandwich syndrome contains all the basic elements of lap-joint wet de-bonding plus the fact that ventilated honeycomb, like the wooden box spar, has the ability to create its own micro-climate—and invariably it is a moist one. Venting of the core by perforation was standard in early honeycomb core stock as it was necessary to remove the volatiles from phenolic type bonding resins. It was also thought that, without venting, differential pressure brought on through a reduction in external pressure at high altitudes would tend to force the skins apart. This effect is indeed present but it is nothing compared with the insidious effect of moisture. The other common elements associated with short-lived honeycomb were acid-etched surfaces and epoxy–polyamide adhesive. This combination appeared to be particularly suitable for sandwich construction as the good filleting behaviour and toughness of the adhesive resulted in very high initial peel strength with acid-etched skins. However, as already pointed out, this combination was singularly short on durability unless provided with a secondary protection and this, by definition, is not practicable with vented honeycomb sandwich.

Apart from the factors already mentioned, i.e. surface treatment, adhesive type, and venting, the almost universal use of clad aluminium for skins was further identified as a likely cause of poor durability.[9,10] As the adoption of clad sheet was a device introduced in the first place solely for improved corrosion protection and longevity (the high-purity aluminium cladding actually imposes a weight penalty) this last item must be regarded as contentious to say the least. A possible explanation for this anomalous situation is given later.

Arising from the generally poor reputation acquired by aluminium sandwich construction in military and airline flying a swing towards the non-metallic sandwich has taken place. GRP, Kevlar, Nomex, carbon-fibre type sandwich structures have increased in use and, whilst there are several reasons for this trend, the freedom from corrosion (there can still be a loss in strength) and the freedom from wet de-bonding are strong arguments in their own right.

Land-based aluminium honeycomb sandwich structure using first-generation materials and procedures have given very little trouble and there are many fully-exposed sandwich radar aerial dishes and similar structures in the UK which seem likely to produce a 20-year life span with only localised repairs and maintenance. Given the latest technology aluminium honeycomb sandwich should be seen as a very durable lightweight trouble-free structure used on the ground.

Returning to the more demanding aircraft sandwich application, the present state-of-the-art aims to correct the several design and manufacturing shortcomings in sandwich technology which have been identified through the experiences described above.

Firstly, attention must be given to the choice of alloy adopted for the honeycomb core. Commercially pure aluminium foil produces honeycomb with excellent—and intrinsic—resistance to corrosion. Higher-strength alloys promise the same structural performance for a lower core density, which is very attractive to the designer. However resistance to corrosion is much lower for these alloys, and, bearing in mind the fact that gauge thickness for aluminium honeycomb is only a few thousandths of an inch (usually less than 100 μm), corrosion, as well as de-bonding, must be seriously considered. The potential corrosion threat is therefore countered by applying a chromate conversion ('Alocrom', 'Alodine', etc.) treatment to the core and, in certain cases, by following this up with a thin, protective, surface coating. Needless to say, the core is no longer deliberately vented.

Turning to the skins, American practice now leans towards unclad sheet for the facings coupled with phosphoric acid-anodise for the pre-bond treatment. The European scene remains, however, much more fragmented at the time of writing. Chrome–sulphuric acid etching and chromic-acid anodising plus various combinations of these and other treatments are still in use and, similarly, there is no general move towards abandoning the use of clad sheet material except where the structure has been designed for absolutely minimum weight, with long-term durability relegated to a lower priority.

The skin-to-core assembly adhesive is still, of course, a modified epoxy (discounting specialised high temperature uses) but the particular adhesive would now be selected more on wetting behaviour and durability, rather than on toughness and peel performance as in the past. It is very likely, too, that the use of a so-called corrosion inhibiting primer (CIP) would be included in the process.

Finally, and notwithstanding all the advances set out above, every effort is now made to achieve complete edge sealing of the panel through the use of gap-filling epoxy and polysulphide type sealants. The effectiveness of the edge sealing is ultimately checked by various forms of reduced-pressure leak tests.

3. PROCESS VARIABLES AND THEIR EFFECTS ON DURABILITY

3.1. Metal Surface Pretreatment

The importance of the treatment of the metal surface, prior to bonding, in the attainment of high strength joints was recognised from the very beginning of structural metal bonding. Most of the technical effort and much of the published literature centred around this aspect of bonding technology. It has already been stated that whilst for many years the aim was to evolve a cost-effective, consistent, production procedure—as measured by the standard lap shear and peel tests—the trend in recent years has been to concentrate more and more on long-term durability. Nevertheless, it appears that once again metal surface pretreatment is paramount, the difference being that the ranking of the options has changed. The various pretreatments employed for aluminium and its alloys have been discussed in detail in Chapters 2 and 5 but will be briefly discussed below from a 'practitioner's' viewpoint.

3.1.1. Metal 'As Received' but Solvent-degreased

The solvent-degreased condition is not normally utilised for serious bonding work but it is included here for completeness and as a 'lowest common denominator'. Even when given a thorough degreasing treatment by solvent wipe or by vapour-bath immersion the surface still remains in a state of low energy, not predisposed towards good bonding. For example, the simple water-break test will show a complete failure to 'wet' the surface and the contact angle will be very high. The nature of the discrete yet protective layer of non-metallic oxide normally present on high-strength aluminium alloys can be gauged by the poor response to electrical spot-welding, the non-wetting of solder, and the failure of mercury to amalgamate or react in any way with the typical aircraft alloys in sheet form, even when thoroughly cleaned. It is therefore clear that any adhesive does not make direct contact and bond to the 'pure' aluminium, but bonds to the oxide layer. It should be noted that intermediate layers of oxide are also present in all the other surface conditions covered in this section, the difference being that in some cases they are thinner and in others they are thicker. Apart from the thickness there are other important differences, namely the degree and the kind of porosity present in the particular oxide layer. An attempt to bring out the salient features of three commonly used aluminium surface treatments has been made in Fig. 10. The typical 'as received' oxide state is not shown but, as indicated above, it is thicker and much more protective than the oxide layer produced by the chrome–sulphuric acid pickle treatment.

In a bonded-joint assembly the simply degreased surface will produce a lap shear strength which is certainly lower but not markedly lower than the shear strength achieved by chemically treated surfaces but the really significant weaknesses lie in the peel behaviour and in the lack of durability. The life expectancy of such joints exposed to a temperate climate with no additional protection would be measured in months and not years.

3.1.2. Mechanically Roughened Surfaces

It seems natural to expect (although Dr N. A. De Bruyne initially argued otherwise[14*]) that the mechanically roughened surface would be favourable to good bonding. Firstly, the real area available for adhesion is somewhat increased. Secondly, the surface energy level is higher because atoms present at asperities (being free of surrounding atoms) are energy-rich compared with those with a normal number of near neighbours. Thirdly, the within-joint stress distribution and the fracture mechanics, once a crack has started, are much more favourable for rough surfaces. The increase in surface energy available is dramatically illustrated by the simple water-wetting test when the high contact angle of the water will change completely to a spreading uniform film when the as-received surface is given a nominal degree of abrasion with carbide paper. The mechanical advantages are also readily demonstrated by simple overlap tests using thin-gauge adherends and a hard, brittle adhesive. In the last respect, however, it has to be admitted that the latest very tough, low modulus, reactive acrylic type of adhesive when fully cured does find it very hard to distinguish between smooth and rough surfaces even when loaded in peel.

Dealing with the more general case of the modified-epoxy structural

* His theory was that with a rough surface air would be trapped at the bottom of the cavity, so preventing entry of adhesive.

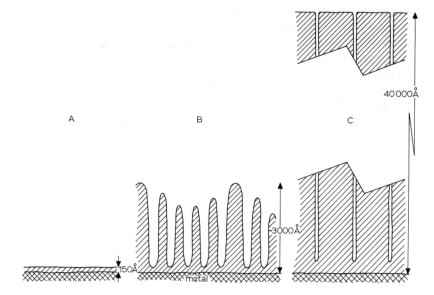

Fig. 10. Schematic diagram of oxidised aluminium surfaces. Macro-roughness and grain boundary etching are not shown.

A, Chrome/sulphuric acid pickle	B, Phosphoric acid anodise	C, Chromic acid anodise (unsealed)
FPL etch OFPL etch (optimised) DTD 915 DEF 03–2/1	Boeing specification 5555, PABST program	Bengough-Stuart, DEF 151 (DTD 910)
An 'active' surface but non-porous to adhesive molecules	Highly porous to most adhesive polymers	Marginally porous to adhesive polymers as normally carried out (porosity can be increased) Highly receptive to polymers in solution— paints, lacquers, etc. Excellent corrosion protection

adhesive, the mechanically roughened surface can be expected to produce joint strengths quite closely approaching (say 90%) those of chemically treated surfaces. Air blasting with fine grit, in particular, gives good and consistent results. Despite these very satisfactory initial joint strength figures, the life expectancy of bonds made to roughened aluminium surfaces does not appear to be significantly increased, compared with solvent-treated surfaces. The time to failure may well be extended somewhat but not to the point sufficient to generate any real increase in confidence in terms of a permanent means of mechanical assembly.

3.1.3. Phosphoric Acid Etch at Ambient Temperatures

The simplest form of chemical treatment available is a swab-on phosphoric acid *etch* used at ambient temperature. Such treatments—there are many proprietary solutions on the market—are particularly attractive for on-site repair work where it is not possible to utilise an elaborate hot-bath process. These phosphoric acid-based solutions are variously known as deoxidisers, conditioners, and etchants—in order of increasing aggression. Applied cold and left for a few minutes a well-etched surface exhibits excellent wetting behaviour results. The initial joint strength figures produced by this treatment do not fully match those of the chrome–sulphuric bath referred to in section 3.1.4 but they are, nevertheless, very acceptable. Unfortunately the durability of joints made with typical epoxy resins is very poor, as is well brought out in Table 1. The Table shows that there is little residual strength after one year of immersion in water in the unstressed condition. The same acid swab treatment used for large area 'in the field' repairs resulted in severe joint de-bonds after a few years of under cover storage in the UK. The use of hot-cure epoxy adhesive in place of the cold-cured system actually used here would have improved the situation but it would not have transformed it. The point mentioned earlier that a thin coat of paint exerts a marked beneficial influence which cannot be simply related to the overall water diffusion behaviour is repeated (the in-the-field repairs mentioned above were not over-painted).

3.1.4. The Chrome–Sulphuric Acid Pickle

The hot chrome–sulphuric acid-etch (or 'pickle') process properly carried out has been shown time and time again in numerous comparative trials to be the treatment capable of giving the greatest *initial* joint

TABLE 1

EFFECT OF PRE-BOND TREATMENT ON JOINT STRENGTH AFTER EXPOSURE TO
WATER AND SALT FOG[a]

Room temperature cured epoxy adhesive on L72 aluminium alloy (16 gauge).

Treatment	Control	Single-lap shear strength (psi)			
		Water immersion		Salt fog environment	
		3 months	1 year	3 months	1 year
Bonded joints unprotected (1 in × ½in lap)					
Chromic acid anodised (40°C)	3360	2650	2880	3430	3120
Chromic acid anodised (32°C)	3680	2710	2040	3370	3380
Phosphoric acid etch/chromate conversion treatment	2730	2290	2300	2440	1580
Phosphoric acid etch only	3760	1370[b]	20[b]	0[b]	0[b]
Bonded joints protected (1 in × ½in lap)					
(Samples as above but protected with epoxy primer plus polyurethane top coat paint scheme.)					
Chromic acid anodised (40°C)	3360	2870	2360	3380	3700
Chromic acid anodised (32°C)	3670	2310	2240	3800	4020
Phosphoric acid etch/chromate conversion treatment	3200	2460	2480	2790	3460
Phosphoric acid etch only	3990	2330	2520	3440	3820

(1000 psi ≡ 6·894 MPa)

[a] All specimens exposed under zero load conditions.

[b] Total joint failure, prior to test, occurred in one or more specimens in this group.

strength irrespective of the adhesive used. Because of the very early recognition of the superior performance attainable from this treatment it is the basis of several national specifications such as the UK DTD915, now Def. Stan. 03-2/1; the USA FPL etch now uprated to the Optimised FPL (OFPL) etch; and the European prEN 2334. It is also frequently selected as the laid down mandatory pretreatment to be used in the making of bonded test specimens in the initial assessment and quality control of the adhesives themselves. As related in section 2, however, the use of this treatment in conjunction with epoxy type adhesives, as distinct from the earlier modified phenolic-based types such as Redux, FM 47 and AF-10, led to the emergence of a clearly identified in-service joint disbond problem frequently associated with attendant corrosion. Accordingly it has become discredited as a treatment in its own right although it still continues to be widely used as one stage in a multistage scheme when the final treatment is generally one or other of the anodic processes. (Bearing in mind the gross changes that take place on the surface when electrolytic oxidation is carried out (Fig. 10) its value as a pre-etch treatment, however effective in its own right, is open to challenge. Indeed, in the case of phosphoric-acid anodising, Ahearn and Venables[11] did in fact find that the prior etch treatment was not necessary in developing the required final oxide morphology.)

The fact that the once pre-eminent chrome–sulphuric prebond treatment has been found wanting and fallen from favour cannot be passed over without some attempt at explanation. Other authors in this book have presented comprehensive and rigorous analyses for the debonding of adhesives from aluminium and other surface-active metals such as magnesium and titanium. However, a practitioner's view, put briefly, is that it is certain that in bonding to these metals we usually have no strong chemical, or valence, bonds directly to the surface. At best we could have some very limited degree of weak chemisorption and/or hydrogen bonding to the oxide surface. However, this appears to be a better situation than when bonding to an inert non-metallic surface such as a well-cured polyester or epoxy laminate. These low energy surfaces do not always give high-strength bonds but, whilst some reduction in strength may be detected in wet environments, debonding does not occur.* The problem with aluminium is that in the presence of water a rapid growth of the oxide layer takes place. The generality of the reaction has been known for a long time but a recent Bell Telephone Laboratories report[12] dealing with integrated circuitry manufacture used Auger electron spectroscopy (Chapter 2) to quantify the oxidation process. On vacuum-deposited aluminium the oxide thickness in air was about 35 Å after several months of exposure. In water the oxide layer grows slowly at room temperature but in warm water the growth is very rapid and an oxide thickness of several thousand Ångstroms is produced within minutes. It is this water-induced oxide growth which the author believes to be the explanation for the particular sensitivity of adhesive bonds to aluminium (and similar reactive metals) when exposed to wet environments. In one sense this oxide growth can be regarded as a form of corrosion but in another sense it is the protective mechanism to which aluminium owes its general serviceability. It should be noted that visually the surface appearance is the same whether the oxide thickness is 30 Å or 3000 Å

* The reasons for interface stability in such cases are discussed in Chapters 1 and 3.

and these subtle changes would not normally be associated with 'corrosion' in the ordinary sense of the word. For these surface changes to take place it has to be accepted that water can arrive at the metal surface by simple diffusion through the adhesive or else by a separate transport process along the interface. There is ample evidence to show that both these mechanisms operate to a greater or lesser extent in all practical adhesives but the former is usually the more important, as discussed in Chapter 3.

By the adoption of selected anodic oxidation processes this problem of instability at the surface can be overcome. If at the same time the resulting stable oxide layer exhibits a degree of controlled porosity and absorption towards the adhesive, and is also strong mechanically, all the ingredients for long-term joint durability are brought together.

Returning to chrome–sulphuric acid pickle, whilst this treatment is no longer a prime choice in its own right it is not obsolete, remaining in use as a prior treatment to anodising and as a standard reference surface, as discussed earlier.

3.1.5. Chromic Acid Anodise

Chromic acid anodising is the standard and long-established protective treatment for aluminium and its alloys widely practised by the aircraft industry worldwide. It is an excellent protective treatment in its own right and it is also used as a first process in complex anticorrosion paint schemes giving very enhanced adhesion and overall performance to the primer component. As a pre-bond treatment for structural adhesives, which of necessity are solventless and far less mobile in nature, chromic acid anodising (CAA) as normally carried out is much less ideal. The characteristic anodically produced oxide layer features a pore size which appears to be too small to accept the typical adhesive polymer molecule to a sufficient degree. Furthermore, a smooth enamel-like macro-surface is frequently produced (depending on alloy type and surface finish) and this translates into a poor peel and cleavage performance when used with high-modulus adhesives. Thus, even though the durability may be good the CAA treatment would be discounted because the initial joint strength can be quite poor. The situation is much improved when CAA is superimposed on to a mechanically roughened or a well-etched surface and a good balance of initial strength and long-term durability can be achieved via this route.

As the original Bengough–Stuart CAA process was intended to

maximise the intrinsic corrosion protective behaviour the overall porosity is, by design, low and the pore diameter is small. Subsequent 'sealing' by immersion in near-boiling water renders this porosity much smaller still. The adhesion behaviour is now very poor although the response to solution systems such as paints is still satisfactory. For use within the context of adhesive bonding a CAA process which maximises the pore diameter would be more rational and indeed this trend has now commenced spurred on by the competition generated by the emerging phosphoric acid anodise process. It is important to note that a CAA surface made more porous by changes to the electrolytic strength, operating voltage, or bath temperature is still very different from the phosphoric acid-anodise surface in that the overall thickness is still far greater (see Fig. 10), and also in that the important protective barrier layer below the pore structure would be expected to remain much thicker.

3.1.6. Phosphoric Acid Anodise

The anodising of aluminium using phosphoric acid electrolyte has been well established in the lithographic printing trade and in metal plating for many years. In both cases the reason for adopting this particular electrolyte is the uniquely porous and absorptive surface which is readily produced. Being thin (for an anodic layer) and porous the performance in terms of corrosion protection is poor and accordingly phosphoric acid anodise (PAA) was, until quite recently, totally ignored by the aviation industry as a treatment for aluminium. More recently, however, the PAA process has been rediscovered and developed in America to meet the specific needs of structural adhesive bonding and a vast amount of work has been undertaken by various manufacturers and by government agencies aimed at characterising and optimising it (see Chapters 2 and 5 for details).

Phosphoric acid anodising produces a thin but very porous oxide layer; this is the result of the mild anodic field conditions and the aggressive nature of the electrolyte. There is ample evidence from work using the electron microscope that the scope for the mechanical engagement of adhesive resin with the metal oxide surface is maximised when this anodic process is used. This feature of high porosity and large pore size appears to be achieved over a wide range of working bath conditions and it indeed appears that the simple mechanical interlock aspect of PAA is outstanding amongst current bonding pretreatments. Work in Japan on the behaviour and adhesion of

Fig. 11. *Typical surface, chromic acid-anodised aluminium alloy.*

powder-type surface coatings concludes that with PAA treatment there is
no obstacle to polymer penetration, which continues to the base of the
pores.[13] The polymer size and the flow behaviour of structural adhesives would be very much in line with the powder coatings used in that
investigation. Also, as discussed in Chapters 2 and 5, the surface
chemistry may play an important role. However, as the PAA oxide
layer is so thin and porous compared with traditional anodic coatings it
is unreasonable to expect good protective properties for it either in the
mechanical (hardness) or in the environmental sense.

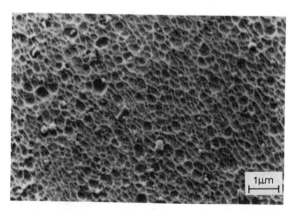

Fig. 12. *Typical surface, phosphoric acid-anodised aluminium alloy.*

The behaviour of adhesive bonds to PAA under short-term laboratory-style accelerated ageing tests has been consistently good—hence the widespread American acceptance. However, in the longer-term real-life situation there must remain some concern that, as with the chrome–sulphuric pickle, the ephemeral nature of the oxide layer might prove to be a weakness.

The characteristic difference between CAA and PAA in terms of the 'mechanical' aspect of adhesion is well brought out in scanning electron micrographs depicting typical anodic treatments carried out on clad alloy sheet, as may be seen from Figs 11 and 12.

3.2 Adhesive Selection—Modified Phenolic or Modified Epoxy?

In general terms, epoxy based adhesives are easier to use in the production environment and are less demanding in terms of process variables. Continued support for the older type of structural adhesives based on phenolic resin stems from the superior environmental behaviour experienced with the simply acid-pickled surface—a subject dealt with in section 3.1.4. The reason for this difference in behaviour between phenolic based and epoxy based adhesives on non-anodised surfaces is a matter of importance. Various explanations have been put forward. One is that as originally used in the original form of Redux (the liquid/powder system) the phenolic resin component could quite legitimately be considered to function in the manner of a primer with all the advantages of a separate, low viscosity, surface-wetting component. Even in the later film form of modified phenolic adhesive the crosslinking phenolic component is still present as a discrete, very mobile, phase.

A second explanation is that substituted phenols as a class can form chelate-linked compounds with metal ions giving a more stable aluminium-to-resin bond. However, as indicated by Fig. 10, it is doubtful if direct contact with metal is ever achieved. A possibility favoured by the author is that the moisture liberated by the polycondensation curing reaction (and necessarily at high temperature) triggers off the rapid surface oxidation reaction already referred to and thereby produces a stabilised surface intimately combined with the curing resin. To achieve this effect a fair thickness of bondline is required so that any attempt to use the same principle via the use of a phenolic-type corrosion-inhibiting primer (CIP) applied under an epoxy based adhesive is not likely to work in the same way. These now popular adhesive bond primers have to be kept very thin or else the peel performance of the overall system is too strongly compromised, whatever their chemical make-up.

4.   THE CLAD ALLOY CONTROVERSY

It has already been stated in section 3, dealing with the problems of poor durability frequently experienced with honeycomb sandwich construction, that the use of clad aluminium sheet was (surprisingly) identified in several American investigations as a feature likely to result in accelerated joint breakdown. Put simply, in laboratory trials with all else equal, bonds made to unclad alloy sheets outperformed—that is, outlived—bonds made to the clad version of the same alloy.

The use of clad alloy sheet has been standard practice within the aircraft industry for very many years. A thin overlay of high purity aluminium is rolled on to the high strength core material (both sides) during manufacture. A weight penalty is involved, but this is accepted as the price to be paid for a significant improvement in the corrosion resistance of the metal and the general serviceability of the airframe.

It seemed strange that a feature designed solely to improve metal surface corrosion behaviour was—in the case of bonded joints—operating in precisely the opposite sense. However strange, the facts were firmly accepted and the consequences—no cladding with bonded assemblies—were soon to appear as mandatory design requirements in several American aircraft in which structural adhesive bonding was employed.

In explanation it was generally put forward that the pure aluminium cladding protected the core material by acting as a 'sacrificial' layer in a corrosive situation and was thereby naturally consumed in the overall process. This is the manner in which the electronegative component in a bi-metallic couple provides its protective influence. However, this action does not occur over the area of the joint as a whole (because there is no 'visible' couple) but only at the termination of the joint where the cut edge of one member exposes the alloyed core to the more electronegative cladding. Here there is formed the classic corrosion cell in which the pure aluminium cladding is preferentially consumed. As the bond is made to the cladding (i.e. the surface) the whole joint assembly is thereby undermined. The theory is sound and the process may well operate to a certain degree in the joint degradation equation. However, it is open to challenge as the major factor because the rate of breakdown should be directly linked to the ionic activity of the electrolyte whereas, as shown in Fig. 8, high humidity and de-ionised water seem frequently to be more damaging than salt solutions.

There is another entirely different explanation for the poor showing of clad material compared to unclad material in experimental programmes where all other variables are made equal. The explanation is simply that the all-important matter of surface treatment does not turn out to be equal despite the fact that the experimental details are the same. It is precisely because the surface treatments—the acid etching or the anodising, etc.—are deliberately made the same that the end result turns out to be dissimilar. The response of the high purity aluminium clad surface to a given bath treatment is significantly different to the response of the non-clad surface which contains other metals, such as copper, zinc, magnesium, etc. In general, a given treatment bath is far less active when acting on the pure aluminium clad surface. If all other things were simply not equal, it is highly likely that the surface treatment obtained on the clad version was milder and less effective than that obtained on the unclad base alloy. With this shortcoming corrected the author believes that the clad versus unclad anomaly is resolved and indeed that the clad aluminium surface is the one to be preferred for the long term serviceability of adhesive bonded light alloy structures.

5. FUTURE TRENDS

Present knowledge gained from extensive work with the electron microscope and modern instrumental surface analysis techniques has resulted in a significant increase in knowledge concerning the real nature of the metal surface involved when making strong structural adhesive bonds. Most of the problem areas are now well understood and the respective solutions have been found. However, in nearly all cases the solution has involved a vast increase in manufacturing complexity and the typical metal surface pretreatment procedure as practised by the aircraft industry is slow, costly, high in technical manpower requirements and generally unattractive when viewed against a stringent economic background. It is to be hoped therefore that in future a high priority is given to attempts to attain the best current standards of adhesive bond performance with simpler methods of surface preparation.

REFERENCES

1. Comparative tests on Redux and Araldite adhesives, *RAE Tech. Note, CHEM 1168* (March 1952).
2. The tropical durability of metal adhesives, *RAE Tech. Note, CHEM 1349* (Feb. 1959).
3. Effect of outdoor exposure on stressed and unstressed bonded joints, *RAE Tech. Report 70081* (May 1970).
4. Kerr, C. and Orman, S. The effect of water on aluminium/epoxide bonds, *City University, London, 7th Conference on Adhesion* (1969).
5. DeLollis, N. J., Theory of adhesion, mechanism of bond failure and mechanism for bond improvement, *Sandia Laboratory Report SC-RR-68-270* (May 1968).
6. Keemar, L. and Strong, R. (BOAC), *Shell Aviation News* No. 366 (1968).
7. McDonald, J. F. (Flying Tiger Line), *Shell Aviation News* No. 445 (1978).
8. Askins, D. R. and Schwartz, H. S., Air Force Materials Lab., Wright-Patterson Base, and University of Dayton Report (Feb. 1978).
9. Green, R. H. *Corrosion resistant adhesive bonding*, Rhor Corporation, Riverside, California Report.
10. Wegmann, R. F. *et al.* Durability of adhesive bonding systems for use in fabrication of an airmobile maintenance shop, *Picatinny Arsenal Report* (1971).
11. Ahearn, J. *et al.*, Development of oxide films on aluminium with phosphoric anodising process, *SAMPE Q.* (Oct. 1980).
12. Chang, C. C. *et al.* (Bell Laboratories), Aluminium oxidation in water, *J. Electrochem. Soc.* (May 1978).
13. Omata, K. *et al.*, Adhesion of paint to anodic oxide films, *Aluminium* (Dec. 1981).
14. De Bruyne, N. A. The extent of contact between glue and adherend, *Aero Research Tech. Note 168* (1956).

Index